A VETERAN GENIUS
OF A COOK

SHOWS YOU HOW TO PREPARE THE RICHEST, MOST LUSCIOUS MEALS YOUR IMAGINATION OR APPETITE COULD DESIRE!

Jennie Grossinger is the celebrity
whose zest for good Jewish food
put Grossinger's famous Catskill resort
on the map and has attracted over 50,000
guests each year. She learned her
traditional recipes in her mother's kitchen;
she is a firm believer in her mother's
maxim, "No one must ever go away hungry!"

All you need for good Jewish cooking
are good ingredients and plenty of them!
Whether familiar or exotic-sounding,
all these enticing foods are easy to prepare
with this delightful,
rewarding cookbook—

Jennie Grossinger's

THE ART OF
JEWISH COOKING

THE ART OF
JEWISH COOKING

JENNIE GROSSINGER
WITH AN INTRODUCTION BY PAUL GROSSINGER

BANTAM BOOKS · TORONTO · NEW YORK · LONDON

THE ART OF JEWISH COOKING
A Bantam Book

PRINTING HISTORY
Random House edition published June 1958
2nd printing .... April 1959   5th printing ..... January 1962
3rd printing ..... January 1960   6th printing .... April 1963
4th printing ...... March 1961   7th printing .... October 1964

Bantam edition published March 1960
2nd printing ...... March 1960
3rd printing .... April 1960
4th printing ...... October 1960
5th printing .... February 1961
Bantam Reference Library edition published March 1962
7th printing ..... October 1962
8th printing ... September 1963
9th printing .... May 1964
Bantam Cookbook Shelf edition published October 1965
11th printing

*Bantam Books are published by Bantam Books, Inc., a subsidiary
of Grosset & Dunlap, Inc. Its trade-mark, consisting of the words
"Bantam Books" and the portrayal of a bantam, is registered in the
United States Patent Office and in other countries. Marca Registrada.
Bantam Books, Inc., 271 Madison Avenue, New York, N. Y. 10016.*

PRINTED IN THE UNITED STATES OF AMERICA

Contents

In this book you will find some recipes which specify the use of either butter or fat. If you wish to cook according to strict dietary observances, you will cook such dishes with fat if they are to accompany meat, with butter if they accompany dairy products.

For observers of the Kashrath: With regard to all recipes having liver as an ingredient, it should be noted that the liver is first to be broiled over an open flame.

Introduction

Since I am neither Josephine McCarthy nor Clementine Paddleford, you may well ask what I am doing writing an introduction to this cookbook.

Well, I've been around kitchens all my life—my father saw to that—and certainly I am an authority on the cooking done by the author of this particular book, Jennie Grossinger. I've been an ardent admirer of her cookery all the years of my life. For she is my mother.

More than most women, Mom had to be a good cook. For if a good cook is the heart of the American home, how much more so is a good cook the heart of the American-plan hotel?

Mom is past the crying-over-the-onions and the dishwashing phases of cooking. Grossinger's—the institution which her smile, her struggles, her love for people, and her way with a saucepan helped to build—now has a pretty good-sized crew of chefs and bottle-washers to handle such matters.

She and her mother, Malke Grossinger, assembled the recipes. My sainted grandmother used to say: "We must never let anyone go away from Grossinger's hungry." That guiding principle together with a characteristic warmth and humanity, a concern for people that went far beyond the ordinary hotel owner-guest relationship are perhaps the two cornerstones that helped build a resort that now sprawls over 1,000 acres of God's good earth and is known in the furthest corners of the globe.

Ever since I can remember, Grossinger women have taken pride in their activities in that most important room of the house, pride and joy in placing on their tables food that was as good as it was plentiful. And it even goes back beyond the days that I can remember.

For it comes down to this. French cuisine may be famous for its Escoffier. Italian for its Alfredo. But Jewish cooking—well, for generations and generations, way back to Sarah, Rebecca and Rachel, the master chef has always been the mistress of the particular tent: Mom.

Mother used to tell me about her early days on New York's crowded East Side, when she came to this country as a seven-year-old immigrant girl. She remembers the hallways of the ancient tenements on Hester and Essex streets not for their dinginess, although there was plenty of that. She recalls the cooking aromas wafted out from beneath apartment doors when she came home from public school.

On the ground floor, perhaps the corned beef and cabbage of an Irish family, heavy and pungent. On the next landing, an Italian family—and the air would be full of the aromatic blending of ripe tomatoes, garlic, olive oil, oregano, and all kinds of other wonderful mouth-watering exotic scents.

But she was happy that the next doorway was hers, for

it would have been impossible for her to pass it by. From behind it, came the delicate but rich Sabbath-preparation smells—of chicken simmering in the pot, of golden soup with feather-light matzo balls floating in it, maybe even the delicious aroma of freshly baked challah.

Food has always been important to the Jewish family, not for itself alone—although it is certainly appreciated—but as a gift from the Lord, which therefore gives it special significance at the most important occasions in life. At a wedding, the bride and groom eat something sweet—for a sweet life together. And the same is true on the Jewish New Year, Rosh Hashonah—this time, for a sweet year.

A Jewish cookbook can be almost considered a history book—a history of 5700 years of happiness and sorrow. Just one instance—the *latke* (pancake), which the wives of the soldiers of the ancient hero Judah Maccabee hurriedly cooked for their men behind the lines, as they united to drive the Syrians out of their land.

Italian cooking is built around pasta, the traditional tomato sauce, and olive oil and special spices; Mexican cooking around hot foods, corns, peppers; French around wine. If you tire of these things, you might as well leave the country. But Jewish cooking, since the kettles of the Hebrews have simmered in every country of the world since the Dispersion many centuries ago—is never monotonous, but thoroughly international in flavor. All it requires is good ingredients and plenty of them.

Jewish cooking is truly a Melting Pot. It has tasted of the spices of Italy, the herbs of the Slavic countries, the tender lamb of Israel, and the goulash of Hungary and middle Europe. Yes, and the potatoes of Ireland, as the Lord Mayor of Dublin testified when he visited Grossinger's. All these things, adapted to the requirements of Jewish law, were brought to the American melting pot by immigrants of the Jewish faith. All these things have made Jewish cooking worldly and sophisticated—though always "home" style—and particularly delicious.

Lin Yutang once wrote something which, it might be

argued, belongs in a Chinese cookbook, but I kind of like it and so I shall borrow it for our purposes here:

"If a man be sensible, and one fine morning, while he is lying in bed count at the tips of his fingers how many things in this life truly give him enjoyment, invariably he will find food is the first one."

I don't know whether you'll agree with Lin Yutang. After all, there are some other pretty important enjoyments in life—children, for instance. But I think you will agree on one point. After you've sampled some of the recipes my mother suggests—all of which have either been tried by, used by, or approved by our Grossinger chefs—I think you'll agree that there is plenty of the kind of enjoyment the words of Lin Yutang suggest in the pages of this book.

So, in Mom's own words: "Hearty appetite!"

Paul Grossinger

In Jewish-style cookery, more so than in most other cuisines, recipes have been handed down for countless generations from mother to daughter. The basic recipes are similar, but each family group has its own variations, depending on its national origin, and always superior to anyone else's, of course.

Jewish dishes are drawn from many lands. Each group has brought with it the traditional foods of its country of origin. From Czechoslovakia come dumplings and the widespread use of poppy seeds; from Poland, the classic pirogen; Germany has supplied gefilte fish and pancakes; goulash was introduced from Hungary; the favorite herring preparations from Holland. In their wanderings, these dishes have been modified and reworked to become a part of what is generically called "Jewish-style cooking." It is interesting to note that this assimilation process is still going on. You will find recipes for pizza, chicken chow mein, chicken Cantonese, egg roll and chile con carne in this book, their presence a testimonial to the ready acceptance of new dishes into an ancient tradition of cooking.

Many of the favorite Jewish dishes have found their places on American menus across the country. People who formerly believed that Jewish cookery was complicated and bizarre have found it well-flavored, easily prepared and delicious.

Jewish holidays usually offer special dishes, often with interesting customs in connection with their observance. Rosh Hashonah (the Day of Judgment) marks the beginning of the New Year and sweet dishes are symbolically served to foretell a happy (and sweet) year. Honey cake is a traditional treat and one of the many different styles of *tzimmes* is a regular part of the holiday's menu.

Yom Kippur (the Day of Atonement) is a fast day, so the meal served before sundown of the preceding day is always substantial, but quite bland—to prevent undue thirst. A typical dinner might include chicken soup, boiled chicken, stewed fruit, sponge cake and tea.

Succoth (the Festival of the Tabernacles) is the week-long festival celebrating the gathering of the harvest, a holiday analogous to Thanksgiving. This is a happy holiday with bountiful meals and an air of celebration. *Kreplach* and stuffed cabbage are two of the favorite dishes for this holiday. Emphasis is customarily placed upon fruits and nuts, the produce of a good harvest.

Chanukah (the Festival of Lights) is an eight-day celebration. Gifts are exchanged in a general atmosphere of gaiety. Potato and cheese pancakes are always prepared for the family, and a selection of rich cakes and cookies are served to guests. Purim is a light-hearted one-day holiday and the specialties are the same today as they have been for countless centuries: *hamentaschen, knishes* and poppy-seed candies.

Passover (the Festival of Deliverance) has its own specialties. Many everyday foods, on the other hand, are specifically forbidden during this period. Leavening (yeasts and other similar ingredients) may not be used during this eight-day period in commemoration of those Jews who, in their haste, were unable to let their bread rise. *Knaidlach, matzo kugel, matzo brie* and *russel borsch* are Passover favorites.

Shavuoth (the Festival of the Torah) is a two-day period which pays respect to the basic laws of the Jews, the Torah. Dairy foods are popular for this holiday: *schav, blintzes, kreplach* and *knishes*. The cycle of holidays ends with *Shavuoth* and is the last until the New Year.

A weekly observance is the Sabbath, which commences at sundown on Friday and ends at sundown on Saturday. No work or cooking is sanctioned, although it is permissible to keep a dish warm, or even to heat it on a previously lit fire, providing it doesn't boil. *Cholent,* a favorite Jewish preparation, is one of the dishes devised for this purpose. An accepted part of the Friday evening meal is gefilte fish, and in many homes it wouldn't be a Friday dinner without *tzimmes*.

Cooking Measurements

DASH, *less than 1/8 teaspoon*

3 TEASPOONS, *1 tablespoon*

2 LIQUID TABLESPOONS, *1 ounce*

4 TABLESPOONS, *1/4 cup*

16 TABLESPOONS, *1 cup*

1 CUP, *1/2 pint*

2 LIQUID CUPS, *1 pound*

16 OUNCES, *1 pound*

4 CUPS, *1 quart*

Oven Temperature Guide (Fahrenheit)

VERY SLOW, *225°*

SLOW, *250° to 300°*

MODERATE, *325° to 375°*

HOT, *400° to 450°*

VERY HOT, *475° and over*

APPETIZERS AND PARTY SNACKS

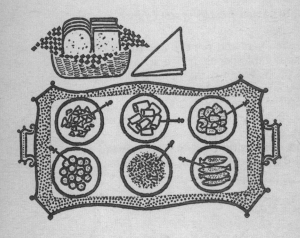

Appetizers play an important part in a Jewish-style meal. They are intended to sharpen your appetite for the courses to follow, and most Jewish families like meals of several courses. Incidentally, many of the appetizers which follow are just about ideal for cocktail parties—bowls of chopped liver served with thinly sliced rye bread; pickled herring cut into small pieces; eggplant spread on pumpernickel. In the FISH section you will find a recipe that has become a great favorite as an appetizer: tiny balls of gefilte fish served with a horseradish dip. Bite-size knishes, blintzes or miniature strudels are popular, too. But above all, remember that appetizers are intended to pique or stimulate your appetite, not satisfy it.

Chopped Chicken Livers

> *1 pound chicken livers*
> *4 tablespoons rendered chicken fat*
> *2 onions, diced*
> *3 hard-cooked egg yolks*
> *1 teaspoon salt*
> *¼ teaspoon freshly ground black pepper*

Wash the livers and remove any discolored spots. Drain.

Heat 2 tablespoons fat in a frying pan; brown the onions in it. Remove the onions. Cook the livers in the fat remaining in the skillet for 10 minutes. You can grind or chop the onions, livers and egg yolks, but be sure you have a smooth mixture. Add the salt, pepper and remaining fat. Mix and taste for seasoning.

Serve cold with crackers as a spread or on lettuce. Serves 6 as an appetizer or 12 as a spread.

Chopped Liver

> *1 pound calf's, beef or chicken livers*
> *2 onions*
> *2 hard-cooked eggs*
> *3 tablespoons rendered chicken fat*
> *1 teaspoon salt*
> *¼ teaspoon freshly ground black pepper*

Wash the liver and combine in a saucepan with 1 onion and water to cover. Bring to a boil and cook over low heat 10 minutes. Drain and discard the onion. Remove skin from liver.

Grind or chop the liver, eggs and remaining onion until smooth. Blend in the fat, salt and pepper. Taste for seasoning.

Serve on lettuce leaves. If you like, garnish with a little chicken fat on each portion. Serves 6 as an appetizer or 12 as a spread.

Vegetarian Chopped Liver

1 cup sliced onion
3 tablespoons butter
2 hard-cooked eggs
2 California sardines (in tomato sauce)
½ cup walnuts
1 teaspoon salt
¼ teaspoon pepper

Cook the onion in the butter for 15 minutes. Place in a chopping bowl and add the eggs, sardines, walnuts, salt and pepper. Chop until very fine. Chill and serve on lettuce with rye bread. Serves 4 as an appetizer or 8 as a spread.

Dairy Liver

3 tablespoons salad oil
1½ pounds mushrooms, sliced
½ cup diced onion
1 hard-cooked egg
1½ teaspoons salt
¼ teaspoon pepper

Heat the oil in a skillet. Cook the mushrooms and onion over medium heat for 10 minutes.

Chop the mushrooms, onion, egg, salt and pepper together until smooth. Chill. Serves 6 as an appetizer or 12 as a spread.

Eggplant Spread

1 medium eggplant
3 tablespoons minced onion
2 tablespoons salad or olive oil
4 tablespoons lemon juice
1½ teaspoons salt
¼ teaspoon pepper
1 teaspoon sugar

Bake the eggplant in a 475° oven until skin turns dark brown. Cool and peel.

Chop the eggplant until very smooth. Stir in the onion, oil, lemon juice, salt, pepper and sugar. Chill. Serve with dark bread and wedges of tomatoes. Serves 4–6.

Calf's Brains Appetizer

1 tablespoon vinegar
2 cups water
2 teaspoons salt
2 calf's brains
¾ cup chopped onions
3 tablespoons salad oil
¼ teaspoon pepper
3 tablespoons lemon juice

Bring the vinegar, water and salt to a boil. Add the brains. Cover and cook over low heat 25 minutes. Drain and cool. Remove the membrane.

Chop the brains, onions, oil, pepper and lemon juice together. Taste for seasoning. Chill. Serves 6.

Chopped Eggs and Onions

> 8 hard-cooked eggs
> ¾ cup chopped onions
> 1 teaspoon salt
> ¼ teaspoon white pepper
> 3 tablespoons rendered chicken fat

Chop the eggs and onions together until very fine. Blend in the salt, pepper and fat. Chill. Arrange on lettuce leaves. Serves 6 as an appetizer or 12 as a spread.

Chickpea "Hot Dogs"

> 1 pound dried chickpeas
> 3 slices white bread, trimmed
> 2 eggs
> ½ teaspoon minced garlic
> ¼ teaspoon diced ground red peppers
> 1¼ teaspoons salt
> ¼ teaspoon pepper
> Fat for deep frying

Wash the chickpeas and soak overnight in cold water. Drain.

Grind the peas and bread together. Mix in the eggs, garlic, red peppers, salt and pepper. Shape into finger-length rolls. Chill for 1 hour.

Heat the fat to 370° in a deep kettle and carefully drop the "hot dogs" into it. Fry until browned. Drain. Serve with mustard or hot chili sauce. Makes about 20.

Petcha

> 2 calf's feet
> 2 onions
> 4 cloves garlic
> 3 quarts water
> 1 tablespoon salt
> ¾ teaspoon black pepper
> 4 hard-cooked eggs

Have the feet chopped up. Pour boiling water over them and scrape with a sharp knife.

Combine the feet, onions, 2 cloves of garlic, the water, salt and pepper in a saucepan. Bring to a boil and cook over low heat 3½ hours. Strain the soup. Cut the meat from the bones and divide among 2 or 3 pie plates. Put the remaining garlic through a press and mix it into the soup. Pour the soup into pie plates. Let it set for ½ hour, then slice the eggs and arrange in the pie plates. Sprinkle with black pepper and chill.

You may serve 12, but if you really like it there's only enough for 6.

Chopped Herring

> 6 fillets of salt herring
> 3 tablespoons chopped onion
> ½ cup chopped apple
> 2 hard-cooked eggs
> 3 tablespoons cider vinegar
> 2 slices white bread, trimmed
> 1 teaspoon sugar
> 2 tablespoons salad oil

Soak the herring in water to cover overnight. Change the water twice. Drain.

Chop the onion, apple, eggs and herring together. Pour the vinegar over the bread and add to the herring with the sugar and oil. Chop until very smooth. Taste for seasoning, adding more vinegar if needed. Chill.

Serves 8 as an appetizer or as many as 24 as a spread.

Jennie's Herring Salad

4 fillets of salt herring
4 scallions
6 radishes
2 tomatoes
2 green peppers
1 cup shredded lettuce
¼ cup cider vinegar
3 tablespoons salad oil
1 teaspoon sugar
¼ teaspoon paprika
¼ teaspoon freshly ground black pepper

Cut the herring into half-inch pieces. Slice the scallions and radishes, cut the tomatoes into quarters or eighths and dice the peppers. Toss these ingredients together with the lettuce, vinegar, oil, sugar, paprika and pepper. Chill and serve as an appetizer. Serves 6.

Pickled Herring

6 fillets of miltz herring
4 onions, sliced thin
1 cup white vinegar
¼ cup water
2 teaspoons sugar
2 teaspoons pickling spice
2 bay leaves
¾ cup sour cream (optional)

Wash herring thoroughly and soak in cold water for 6 hours; drain.

Cut the herring into 2-inch pieces. In a glass jar or bowl, arrange alternate layers of herring and onions. Bring the vinegar, water, sugar, pickling spice and bay leaves to a boil. Cool slightly and pour over the herring. Cover tightly and shake. Place in the refrigerator for 48 hours before serving. For dairy meals, the liquid may be mixed with the sour cream before or after pickling. This may be kept in refrigerator for a week. Serves 6–12.

Fried Herring

> 4 fillets of salt herring
> ⅓ cup dry bread crumbs
> ¼ cup flour
> 1 egg
> 2 tablespoons light cream
> 4 tablespoons unsalted butter

Soak the herring in water to cover overnight. Change the water once. Drain.

Mix the bread crumbs and flour on a piece of waxed paper. Beat the egg and cream together in a shallow bowl. Dip the herring in the bread-crumb mixture, the egg mixture and then again in the bread-crumb mixture.

Melt 2 tablespoons butter in a skillet. Brown the herring on both sides, adding butter as needed. Serves 4–6.

Herring Forshmak

> 4 fillets of salt herring
> 4 tablespoons unsalted butter
> 2 onions, diced
> 4 slices white bread

1 apple, peeled and diced
1 cup sour cream
2 tablespoons bread crumbs

Soak the herring in water to cover overnight. Change the water twice. Drain.

Melt the butter in a skillet. Brown the onions in it. Coarsely chop the herring, bread, apple, sour cream and browned onions. Turn into a buttered baking dish and sprinkle with the crumbs.

Bake in a 425° oven 25 minutes. Serve hot. Serves 6.

Baked Herring

4 fillets of salt herring
½ cup milk
⅓ cup flour
4 tablespoons unsalted butter
3 onions, sliced

Soak the herring in water to cover overnight. Change the water once. Drain.

Dip the herring first in the milk and then in the flour (reserving the milk). Melt 2 tablespoons butter in a baking dish and arrange the herring in it. Cover with the sliced onions. Dot with remaining butter and add reserved milk.

Bake in a 375° oven 30 minutes. Serves 4–6.

Baked Herring and Potatoes

6 fillets of salt herring
3 tablespoons butter
4 boiled potatoes, peeled and sliced
2 onions, sliced
2 tablespoons dry bread crumbs

Soak the herring overnight in cold water to cover. Drain and cut into 1-inch pieces.

Melt 1 tablespoon butter in a baking dish. Arrange layers of the potatoes, herring and onions in it, starting and ending with the potatoes. Sprinkle with the bread crumbs and dot with remaining butter.

Bake in a 400° oven 25 minutes. Serves 6 as an appetizer.

Fish Sticks

3 fillets of sole
¾ cup flour
2 teaspoons salt
¼ teaspoon pepper
3 eggs, beaten
¾ cup dry bread crumbs
Fat for deep frying

Cut the sole into 1-inch strips. Add the salt and pepper to the flour and roll the fish strips in the mixture. Dip fish strips in the eggs and then roll them in the bread crumbs.

Heat the fat to 380° and fry the sticks until browned. Drain and serve with catchup or tartar sauce. Makes 24 pieces.

Fish Balls

2 pounds salmon
1 tablespoon salt
½ teaspoon pepper
¼ pound butter
¼ cup flour
¼ cup heavy cream
½ cup oil for frying

Grind or chop the fish very fine and add the salt and pepper. Cream the butter and flour and work mixture into the fish. Beat until very smooth and stir in the cream. Shape into 1-inch balls.

Heat the oil in a skillet and fry the balls until browned on all sides. Serve with a Hollandaise sauce. Serves 6.

Egg Roll

PANCAKES:
> 2 eggs
> ½ cup water
> ½ teaspoon salt
> ½ cup sifted flour
> 1 tablespoon oil

Beat the eggs, water and salt together. Beat in the flour. Heat a 7-inch skillet with a little oil and pour a little of the batter into it to make a thin pancake. Cook until browned on the bottom. Remove, browned side up, and stack while preparing the filling.

FILLING:
> 1 cup sliced celery
> ½ cup sliced onions
> ¼ cup sliced scallions
> 1 cup Chinese or green cabbage
> 2 tablespoons oil
> 1½ teaspoons salt
> ¼ teaspoon pepper
> ½ cup flaked tuna or julienne chicken

Cook the celery, onions, scallions and cabbage in the oil for 5 minutes, stirring frequently. Stir in the salt, pepper and tuna or chicken. Cool.

Place a heaping tablespoon of the filling at one end of each pancake and roll up, tucking opposite end in. Seal with a little beaten egg and chill. Fry in deep fat heated to 380° until browned.

Serve with hot mustard and duck sauce. Makes about 5.

Chinese Sweet-and-Sour Meat Balls

1½ pounds ground beef
2 teaspoons salt
½ teaspoon pepper
2 eggs
4 tablespoons flour
1 cup oil
3 green peppers, diced
1¼ cups beef broth
1½ cups pineapple chunks
3 tablespoons cornstarch
1 tablespoon soy sauce
½ cup cider vinegar
½ cup sugar

Mix together the beef, 1 teaspoon salt and the pepper. Shape into 24 balls. Mix together the eggs, flour and remaining salt. Dip the meat balls in this batter.

Heat the oil in a skillet and brown the meat balls over low heat. Remove meat balls and reserve. Pour off all but 2 tablespoons oil. Add the green peppers, broth and pineapple. Cook over low heat 10 minutes. Mix together the cornstarch, soy sauce, vinegar and sugar. Stir into the mixture and cook until thickened. Return meat balls and cook over low heat 5 minutes. Serves 6.

Pizza

> 1 package hot-roll mix
> 4 tablespoons olive oil
> ½ cup minced onions
> 1 can (8 ounce) tomato sauce
> 1 can (6 ounce) tomato paste
> 1 clove garlic, minced
> ¼ teaspoon oregano
> 1 teaspoon salt
> ½ pound Swiss cheese, grated
> 10 anchovies, chopped
> 2 tablespoons minced parsley
> ¼ cup Dry Gouda

Prepare the roll-mix as package directs and let rise.

Heat 2 tablespoons olive oil in a saucepan; cook the onions in it for 10 minutes. Add the tomato sauce, tomato paste, garlic, oregano and salt. Cover and cook over low heat 15 minutes.

When dough has risen, divide in 2 and roll out to fit 2 jelly-roll pans (11 x 16 inches). Brush with the remaining olive oil. Spread with half the Swiss and cover with the sauce. Cover with the remaining Swiss. Sprinkle with the anchovies, parsley and Dry Gouda.

Bake in a 450° oven 20 minutes. Cut into squares and serve hot.

You can make cocktail pizzas by cutting the dough into 3-inch circles before covering with the ingredients. Makes 8–10 as appetizer, 20–24 as cocktail snacks.

Submarine Sandwich

Cut a long French or Italian bread in half lengthwise and spread one side with mustard and relish. Arrange slices of corned beef, pastrami, tongue and salami on it. Cover with other half of bread. This may be cut into 2-inch slices to serve with cocktails or eaten as a whole sandwich.

Grilled American Cheese and Tomato Sandwich

For each sandwich, butter 2 slices of white bread. Spread the unbuttered sides with prepared mustard and place 2 slices of American cheese and two slices of tomato between the bread, buttered sides out. Brown in a skillet on both sides or grill until cheese melts.

No Jewish-style meal is complete without a soup. Whether it's a clear (but rich) chicken soup or a thick barley and bean, it has a place on the menu. Many soups are hearty and nourishing enough to be simple meals in themselves. Homemade soups are delicious, easy to prepare and do not take a great deal of time, as some people think. You'll find it worth-while to prepare a large quantity, particularly if you have a freezer and can freeze excess amounts.

DAIRY SOUPS

Tomato Soup

> 3 cups canned tomatoes
> 1 cup water
> 4 tablespoons grated onion
> 1½ teaspoons salt
> 2 teaspoons sugar
> 1 cup scalded milk
> ⅔ cup cooked rice

Over low heat, cook the tomatoes, water, onion, salt and sugar for 30 minutes. Force through a food mill. Stir in the milk and rice. Serve hot. Serves 6.

Spinach Soup

> 2 packages frozen spinach
> ½ cup water
> 3 tablespoons butter
> 3 tablespoons minced onion
> 3 tablespoons flour
> 6 cups milk
> 1½ teaspoons salt
> ¼ teaspoon pepper
> Dash nutmeg

Cook the spinach in the water for 5 minutes. Force it through a food mill.

Melt the butter in a saucepan and cook the onion in it for 5 minutes. Blend in the flour and gradually add the milk. Stir in the spinach, salt, pepper and nutmeg. Cook over low heat 15 minutes. Serves 6–8.

Scotch Barley Soup

4 carrots, grated
3 onions, chopped
2 parsnips, diced
4 tablespoons butter
2 quarts water
1 cup pearl barley
2 teaspoons salt
½ teaspoon pepper
2 tablespoons chopped parsley

Cook the carrots, onions and parsnips in the butter for 15 minutes. Add the water; bring to a boil and stir in the barley, salt and pepper. Cook over low heat 1½ hours. Sprinkle with the parsley. Serves 6.

Mushroom-Barley Soup

6 dried mushrooms
3 tablespoons pearl barley
2 quarts water
2 teaspoons salt
¼ teaspoon pepper
2 onions, diced
2 tablespoons butter
2 tablespoons flour
¾ cup milk

Wash the mushrooms and soak in cold water 10 minutes. Slice fine.

Combine the mushrooms, barley, water, salt and pepper in a saucepan. Cook over low heat 1 hour. Brown the onions in the butter and add to the soup. Cook 30 minutes.

Mix the flour in the milk and add to the soup. Cook 15 minutes. Serves 8.

Vegetable Soup

    2 onions, diced
    3 tablespoons butter
    2 quarts water
    2 carrots, grated
    4 cups mixed vegetables (green peas, green beans,
       lima beans, cabbage)
    2½ teaspoons salt
    ¼ teaspoon pepper
    3 potatoes, peeled and diced
    2 tomatoes, diced
    2 sprigs dill
    1 cup milk

Brown the onions in the butter in a deep saucepan and add the water, carrots, vegetables, salt and pepper. Cover and cook over low heat 10 minutes. Add the potatoes, tomatoes and dill. Cook 25 minutes. Stir in the milk and serve hot. Serves 8.

Split Pea Soup

    2 cups split peas
    2½ quarts water
    2 onions, diced
    1 carrot, grated
    2 stalks celery, sliced
    2 sprigs parsley

2 sprigs dill
3 potatoes, peeled and diced
2½ teaspoons salt
½ teaspoon pepper
1 cup milk
1 tablespoon butter

Wash the peas and combine with the water, onions, carrot, celery, parsley and dill. Cover; bring to a boil and cook over low heat 2½ hours. Add the potatoes, salt and pepper. Cook 20 minutes. Stir in the milk and butter, bring to boiling point and serve. Serves 8.

Potato Soup

3 tablespoons butter
1 cup diced onions
3 cups cubed potatoes
1 grated carrot
3 cups water
2 teaspoons salt
½ teaspoon pepper
1 teaspoon caraway seeds (optional)
2 tablespoons farina
3 cups milk
3 tablespoons minced parsley
½ cup sour cream

Melt the butter in a saucepan and brown the onions. Add the potatoes, carrot, water, salt, pepper and caraway seeds. Bring to a boil and stir in the farina. Cook over low heat 20 minutes. Stir in the milk and parsley and bring to boiling point. Garnish with the sour cream. Serves 6–8.

Cream of Corn Soup

¼ cup minced onion
3 tablespoons butter
2 tablespoons flour
3 cups milk
1 can (#2) cream-style corn
2 teaspoons salt
⅛ teaspoon pepper
2 teaspoons minced pimento

Cook the onion in the butter for 10 minutes. Sprinkle with the flour and gradually add the milk, stirring constantly until mixture reaches the boiling point. Add the corn, salt, pepper and pimento. Cook over low heat 10 minutes. Serves 6.

Cold Schav

1 pound schav (sour grass), washed and shredded
2 onions, minced
2 quarts water
2 teaspoons salt
1 tablespoon lemon juice
4 tablespoons sugar
2 eggs
1 cup sour cream

Combine the schav, onions, water and salt in a saucepan. Bring to a boil and cook over low heat 45 minutes. Add the lemon juice and sugar. Cook 10 minutes longer and taste to correct seasoning.

Beat the eggs in a bowl. Gradually add the soup, stirring steadily to prevent curdling. Chill. Garnish with the sour cream. Makes about 1½ quarts.

Cold Borsch

10 large beets, peeled and grated
2½ quarts water
1 onion, minced
2½ teaspoons salt
2 tablespoons sugar
¼ cup lemon juice
2 eggs
1 cup sour cream

Combine the beets, water, onion and salt in a saucepan. Bring to a boil and cook over low heat 1 hour. Add sugar and lemon juice. Cook 10 minutes and taste to correct seasoning.

Beat the eggs in a bowl. Gradually add the soup, stirring steadily to prevent curdling. Chill and serve with boiled potatoes. Garnish with sour cream. Makes about 2 quarts.

Cabbage Soup

4 pounds cabbage, shredded
2 onions, chopped
4 tablespoons butter
2 tablespoons flour
4 cups water
2 cups canned tomatoes
2 teaspoons salt
½ teaspoon pepper
2 tablespoons sugar
2 tablespoons lemon juice
1 teaspoon caraway seeds
1 cup sour cream

Cook the cabbage and onions in the butter for 15 minutes. Sprinkle with the flour and gradually add the water, stirring constantly until mixture reaches the boiling point. Add the tomatoes, salt, pepper, sugar, lemon juice and caraway seeds. Cook over low heat 1 hour. Correct seasoning. Garnish with sour cream. Serves 6–8.

Cottage Cheese Soup

1 onion, chopped
1 stalk celery, chopped
2 green peppers, chopped
3 tablespoons butter
1½ teaspoons salt
¼ teaspoon pepper
½ teaspoon paprika
3 cups water
3 cups milk
½ cup light cream
1 cup cottage cheese

Cook the onion, celery and green peppers in the butter for 15 minutes. Add the salt, pepper, paprika, water and milk. Cover and cook over low heat 1 hour. Just before serving, add the cream and cheese. Heat but don't boil. Serves 8.

Cold Fruit Soup

1 cup pitted plums
1 cup pitted sour red cherries
1 cup sliced peaches
6 cups water
¼ cup sugar

½ teaspoon salt
1 teaspoon cinnamon
2 tablespoons cornstarch
6 tablespoons sour cream

Combine the fruit, water, sugar, salt and cinnamon in a saucepan. Bring to a boil and cook over low heat 15 minutes. Force through a food mill.

Mix the cornstarch with a little water and stir into the fruit mixture. Cook over low heat 10 minutes, stirring frequently. Chill and garnish with sour cream. Serves 6.

MEAT SOUPS

Chicken Soup

1 soup chicken
Chicken feet
3½ quarts water
2 onions
1 tablespoon salt
2 carrots
3 stalks celery
1 parsley root
2 sprigs dill
3 sprigs parsley

Clean the chicken and feet thoroughly. (The feet add strength to the soup, so use as many as you can get.) Combine in a deep saucepan with the water and onions. Bring to a boil and cook over medium heat 1½ hours. Add remaining ingredients. Cover and cook over low heat 1 hour longer, or until chicken is tender. Remove chicken and strain soup.

Makes about 2–2½ quarts soup. Use the chicken in other dishes or serve with the soup.

Potage à la Reine

3 tablespoons chicken fat
3 tablespoons minced onion
¼ cup flour
4 cups chicken broth
½ cup shredded chicken

Heat the chicken fat and cook the onion in it 10 minutes. Add the flour and then gradually the broth, stirring constantly until mixture reaches the boiling point. Add the chicken and cook over low heat 10 minutes. Serves 4–5.

Flanken Soup

3 pounds plate flank
Beef bones
3½ quarts water
1 onion
1 tablespoon salt
Soup greens
1 bay leaf
¼ teaspoon peppercorns

Combine the beef, bones and water in a deep saucepan. Bring to a boil and skim. Add the onion, salt, soup greens, bay leaf and peppercorns. Cover loosely and cook over low heat 2 hours, or until meat is tender. Remove beef and strain the soup.

Makes about 2½ quarts soup. Serve the beef with horseradish. Serves 8–10.

Barley-Bean Soup

1½ cups dried lima beans
2 pounds plate flank
2½ quarts water
¼ cup pearl barley
2 onions, diced
2 teaspoons salt
½ teaspoon pepper
2 tablespoons chopped parsley

Soak the beans overnight in water to cover. Drain.

Cook the meat in the water for 45 minutes. Add the barley, onions and beans. Cook over low heat 1½ hours. Add the salt, pepper and parsley. Cook 10 minutes, or until the meat and beans are tender. Serves 6–8.

Lentil Soup

2 cups lentils
2½ quarts water
2 onions, diced
2 tablespoons fat
2 carrots, diced
1 tablespoon salt
¼ teaspoon pepper
1 bay leaf
4 frankfurters, sliced

Wash and drain the lentils. Combine with the water; bring to a boil and cook over medium heat 1 hour.

Brown the onions in the fat and add to the lentils with the carrots, salt, pepper and bay leaf. Cook over low heat

2 hours. Discard the bay leaf and rub mixture through a food mill. Return to the saucepan. Add the frankfurters and cook over low heat 10 minutes. Serves 6–8.

Meat Split Pea Soup

> 3 quarts water
> 2 cups split peas, washed and drained
> 2 pounds plate flank
> Beef bones
> 1 tablespoon salt
> ½ teaspoon pepper
> 2 carrots, grated
> 2 onions, diced

Combine the water and peas in a saucepan. Bring to a boil and cook over low heat 1 hour. Add the flank, bones, salt, pepper, carrots and onions. Cover and cook over low heat 2 hours or until meat is tender. Remove meat and bones.

Rub the soup through a sieve. Serve the soup with pieces of meat as garnish. Makes about 2 quarts of soup. Serves 8–10.

Meat Borsch

> 3 quarts water
> 2 pounds brisket
> Beef bones
> 8 beets, grated
> 2 onions, diced
> 2 cloves garlic, minced
> 1 tablespoon salt

3 tablespoons brown sugar
⅓ cup lemon juice
2 eggs, beaten

Combine the water, meat and bones in a deep saucepan. Bring to a boil and skim. Add the beets, onions, garlic and salt. Cover and cook over medium heat 2 hours. Add the brown sugar and lemon juice. Cook 30 minutes. Taste to correct seasoning if necessary.

Beat the eggs in a bowl. Gradually add a little hot soup, beating steadily to prevent curdling. Return to saucepan. Serve with pieces of meat as garnish. Serves 8–10.

Cabbage Borsch

2 pounds brisket
Beef bones
2 quarts water
2 onions, diced
3 cups canned tomatoes
3 pounds cabbage, coarsely shredded
2 teaspoons salt
½ teaspoon pepper
¼ cup lemon juice
2 tablespoons sugar
3 tablespoons seedless raisins

Combine the brisket, bones and water in a deep saucepan. Bring to a boil and skim. Add the onions and tomatoes. Cover and cook over low heat 1 hour. Add the cabbage, salt and pepper. Cook 1 hour. Stir in the lemon juice, sugar and raisins. Cook 20 minutes. Taste to correct seasoning if necessary. Serve with meat as garnish. Serves 6–8.

Sauerkraut Soup

> 1 tablespoon fat
> 2 pounds flank
> 1½ cups minced onions
> Beef bones
> 1 pound sauerkraut
> 2 quarts water
> 2 teaspoons salt
> ½ teaspoon pepper
> 1½ cups peeled and cubed potatoes

Heat the fat in a saucepan. Brown the flank and onions in it. Add the bones, sauerkraut, water, salt and pepper. Bring to a boil and cook over low heat 2 hours. Add the potatoes and cook 20 minutes longer. Serve with the meat as garnish. Serves 6–8.

SOUP ACCOMPANIMENTS

Including All Kinds of Knaidlach

In the Jewish cuisine, a soup's importance is evaluated by its thickness. A thin soup is not completely disregarded, but to most people it is of little importance. But a thick, heavy soup—that's an entirely different matter! And when the soup is so thick with soup accompaniments that the liquid can hardly be seen, that is when the cook receives the compliments.

Mandlen, kreplach, farfel and *kasha* are favorite items to serve with soup. It's an amusing idea to try a variety of accompaniments at the same time because, after all, the more the merrier.

You'll find additional suggestions for soup accompaniments in the chapter called NOODLES AND KREPLACH.

Fried Soup Noodles

> 1 cup flour
> ¼ teaspoon salt
> 1 egg
> Fat for frying

Make a well in the center of the flour and drop the salt and egg into it. Work in the flour and knead until smooth and elastic. Cover with a bowl for 20 minutes, then roll out as thin as possible on a lightly floured surface. Stretch the dough gently until it is even thinner. Let dry for 20 minutes, then fold dough in half. Cut in circles with a small melon-ball cutter or thimble.

Heat the fat to 375° and drop the dough into it. Fry until lightly browned. Drain on paper towels and serve in beef or chicken soup. Serves 8.

Farfel

> 2 cups flour
> ½ teaspoon salt
> 2 eggs

Combine the flour and salt. Work in the eggs until a stiff dough is formed. Roll between the hands into narrow strips. Let dry until stiff enough to grate, then grate or chop into pieces the size of barley. Spread out on a towel or board and let dry again.

Store in tightly covered container. To use, cook for 10 minutes in boiling salted water or soup. Serve to your taste, according to your preference for thick or thin soup.

Soup Nockerl

>   2 eggs
>   ¼ cup water
>   1 teaspoon salt
>   ½ cup sifted flour
>   ½ teaspoon baking powder

Beat the eggs, water and salt together. Mix in the flour and baking powder.

Drop by the teaspoon into boiling salted water. Cook until they rise to the surface, about 5 minutes. Drain. Serve to your taste, according to your preference for thick or thin soup.

Mandlen

>   1½ cups sifted flour
>   ½ teaspoon salt
>   ¾ teaspoon baking powder
>   2 eggs
>   1½ tablespoons salad oil

Sift the flour, salt and baking powder into a bowl. Beat the eggs and oil together and blend into the flour, kneading until smooth.

Roll pieces of dough into ¼-inch-thick strips. Cut in ½-inch pieces. Arrange on a greased baking sheet.

Bake in a 375° oven 15 minutes. Shake the pan occasionally. Cool any left over and store in a closed jar. Serve to your taste, according to your preference for thick or thin soup.

Liver Balls

> 3 tablespoons minced onion
> 2 teaspoons melted chicken fat
> ½ pound chicken livers
> 1 teaspoon salt
> ⅛ teaspoon pepper
> 2 tablespoons potato flour
> 1 egg yolk
> 1 egg white, stiffly beaten

Brown the onion in the fat. Chop or grind the raw livers with the onion. Stir in the salt, pepper, potato flour and egg yolk. Fold in the egg white.

Drop by the teaspoon into boiling soup or salted water. Cook 15 minutes or until they rise to the surface. Serves 8–10.

Mandlen (Passover)

> 3 eggs
> 1 teaspoon salt
> ½ cup cake meal
> 2 tablespoons potato flour
> Fat for deep frying

Beat the eggs and salt together. Stir in the cake meal and potato flour.

Heat the fat to 380°. Drop the mixture by the teaspoon into the fat. Fry until browned. Drain and serve hot. If you want to prepare them ahead of time, crisp them in a hot oven before serving. Serves 6–8.

Knaidlach (Passover)

> 2 eggs
> 4 tablespoons melted chicken fat
> ⅓ cup cold water
> 1 teaspoon salt
> 1 cup matzo meal

Beat the eggs, fat, water and salt together. Stir in the matzo meal, adding just enough to make a stiff batter. Chill one hour.

Form into balls and cook for 30 minutes in boiling soup or salted water. Makes about 18.

Fluffy Knaidlach (Passover)

> 3 egg yolks
> 1 teaspoon salt
> ¾ cup matzo meal
> 3 egg whites

Beat the egg yolks and salt together. Stir in the matzo meal and chill 1 hour. Beat the egg whites until stiff but not dry and fold into the matzo-meal mixture. Form into 18 balls. Cook in boiling soup or salted water for 20 minutes.

Cheese Knaidlach (Passover)

> 2 cups pot cheese
> 2 egg yolks, beaten
> ½ teaspoon salt
> 4 tablespoons matzo meal
> 3 tablespoons melted butter
> 2 tablespoons sugar (optional)
> 2 egg whites, stiffly beaten

Force the cheese through a food mill. Stir in the egg yolks, salt, matzo meal and butter. Add the sugar if you want to serve the knaidlach as a dessert, but omit if they are to be served in a dairy soup. Fold in egg whites and chill 30 minutes. Moisten hands; form mixture into 2-inch balls. Cook in boiling salted water 20 minutes or until they rise to the top. Drain. Serve in soup or as a dessert with sugar, cinnamon and sour cream. Makes about 18.

Pareve Knaidlach (Passover)

2 egg yolks
½ teaspoon salt
2 egg whites, stiffly beaten
½ cup matzo meal

Beat the egg yolks and salt until thick. Fold into the egg whites, then gradually fold in the matzo meal. Chill 1 hour. Moisten hands; shape mixture into ½-inch balls. Cook in boiling salted water 20 minutes. Serve in dairy or meat soups. Makes about 16.

Mavim's Knaidlach (Passover)

4 egg yolks
1 teaspoon salt
Dash cayenne pepper
2 teaspoons grated onion
2 tablespoons melted chicken fat
4 egg whites, stiffly beaten
¾ cup matzo meal

Beat together the egg yolks, salt, cayenne pepper, onion and fat until creamy; fold into the egg whites. Gradually fold the matzo meal into the egg mixture. Chill 1 hour.

Moisten hands and shape the mixture into ½-inch balls. Cook in a covered saucepan 30 minutes. Makes about 24.

Potato Knaidlach (Passover)

2 eggs
1½ teaspoons salt
2 tablespoons grated onion
⅓ cup potato flour
3 tablespoons matzo meal
4 cups grated drained potatoes

Beat the eggs, salt and onion together. Stir in the potato flour, matzo meal and potatoes. Shape into 1½-inch balls. Cook in salted water 20 minutes or until they rise to the top. Drain. May be served with meat dishes, too. Makes about 18.

Marrow Balls (Passover)

Large marrow bone
1 egg
2 teaspoons grated onion
1 tablespoon minced parsley
1 teaspoon salt
⅓ cup matzo meal

Have the butcher crack the bone. Carefully remove the marrow (there should be 2 to 3 tablespoons). Cream the marrow, then add the egg. Beat until mixture thickens. Stir in the onion, parsley and salt. Add the matzo meal, a tablespoon at a time until mixture is thick. You may not need all the meal. Chill 1 hour. Moisten hands and shape the mixture into very small balls. Cook in soup for 15 minutes. Makes about 24.

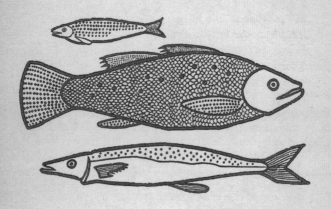

Jewish cookery stresses the value of fresh-water in preference to salt-water fish. The reasons are not difficult to ascertain, for fish cookery in the Jewish fashion is inherited largely from Poland, Czechoslovakia and Hungary, all inland countries which do not have access to salt water. Fish dishes, therefore, were created from what was available—carp, whitefish and pike, the locally caught fish. (If not available, use any firm-fleshed white-meat fish.)

There are many interesting and attractive ways of preparing dishes that keep well and provide delicious main courses for dairy meals or unusual first courses for a substantial dinner. Thanks to freezing techniques, almost all varieties of fish are now available throughout the country. Calorie counters will find fish appetizing and satisfying although low in calories and fat.

Boiled Carp

>*4 onions, sliced*
>*2 beets, peeled and sliced*
>*3 cups water*
>*2 teaspoons salt*
>*½ teaspoon pepper*
>*6 slices carp*

Combine the onions, beets, water, salt and pepper in a deep skillet. Place the fish in it. Cover, bring to a boil and cook over low heat 45 minutes, removing the cover after 30 minutes. Serve hot or chilled. Serves 6.

Scharfe Fish

>*3 onions, sliced*
>*3 carrots, sliced*
>*2 stalks celery, sliced*
>*3 cups water*
>*2 teaspoons salt*
>*¼ teaspoon pepper*
>*6 slices pike*
>*2 egg yokes*
>*2 tablespoons lemon juice*
>*2 tablespoons chopped parsley*

Combine the onions, carrots, celery, water, salt and pepper in a saucepan. Bring to a boil and add the fish. Cover and cook over low heat 30 minutes. Transfer fish to a platter and strain the stock.

Beat the egg yolks and lemon juice in a small saucepan. Gradually add the stock, mixing steadily. Cook over low heat, stirring steadily until it thickens, but do not let boil. Pour over the fish. Sprinkle with the parsley. Serve hot or cold. Serves 6.

Gefilte Fish

> 2 *pounds whitefish*
> 2 *pounds pike*
> 2 *pounds winter carp*
> 5 *onions*
> 2 *quarts water*
> 4 *teaspoons salt*
> 1½ *teaspoons pepper*
> 3 *eggs*
> ¾ *cup ice water*
> ½ *teaspoon sugar*
> 3 *tablespoons matzo or cracker meal*
> 3 *carrots, sliced*

Have the fish filleted but reserve the head, skin and bones. You may use any combination of fresh-water fish although this combination is most popular.

Combine head, skin, bones, and 4 sliced onions with 1 quart of water, 2 teaspoons salt and ¾ teaspoon pepper. Cook over high heat while preparing the fish.

Grind the fish and remaining onion. Place in a chopping bowl and add the eggs, water, sugar, meal and remaining salt and pepper. Chop until very fine; this is important for fluffy fish. Moisten hands; shape mixture into balls. Carefully drop into fish stock. Add the carrots. Cover loosely and cook over low heat 1½ hours. Remove the cover for the last ½ hour. Taste to correct seasoning. Cool the fish slightly before removing to a bowl or platter. Strain the stock over it, and arrange carrots around it. Chill. Serve with horseradish.

You can serve 12 people generously but the fish keeps for a few days so you might as well make this amount even for fewer people.

Baked Gefilte Fish

> *3 pounds halibut or pike fillets*
> *1 onion*
> *2 slices white bread soaked in water*
> *½ cup water*
> *2½ teaspoons salt*
> *½ teaspoon pepper*
> *1 egg*
> *2 tablespoons salad oil*
> *2 onions, sliced*
> *1 green pepper, diced*
> *1 cup canned tomato sauce*

Grind the fish and onion. Transfer to a chopping bowl and add the soaked bread squeezed dry, 1½ teaspoons salt, ¼ teaspoon pepper and the egg. Chop until fine and smooth. Shape into 12 balls.

Combine the oil, sliced onions, green pepper, tomato sauce and remaining salt and pepper in a baking dish. Arrange the balls in it.

Bake in a 325° oven 45 minutes. Baste frequently. Serves 6.

Lemon Fish

> *¼ cup salad oil*
> *2 cups sliced onions*
> *6 slices pike, whitefish or salmon*
> *2 teaspoons salt*
> *½ teaspoon pepper*
> *2 tomatoes, diced*
> *2 lemons, sliced thin*
> *½ cup water*

1 tablespoon cider vinegar
2 teaspoons sugar
1 bay leaf

Heat the oil in a deep skillet. Brown the onions in it. Arrange the fish over the onions and sprinkle with the salt and pepper. Add the tomatoes, lemon slices, water, vinegar, sugar and bay leaf. Cover and cook over low heat 35 minutes. Remove bay leaf. Serve hot or cold. Serves 6.

Fish with Sour-Cream Sauce

3 tablespoons butter
1 cup chopped onions
4 slices whitefish or pike
1 teaspoon paprika
1½ teaspoons salt
¼ teaspoon pepper
1 cup water
½ cup sour cream

Melt the butter in a skillet. Brown the onions, then arrange the fish over them. Sprinkle with the paprika, salt and pepper. Add the water. Cook over low heat 45 minutes. Stir the sour cream into the sauce. Taste to correct seasoning and serve hot. Serves 4.

Fried Fish

¼ cup white corn meal
¼ cup flour
2 teaspoons salt
½ teaspoon pepper

6 fillets or slices of fish
4 tablespoons salad oil
2 onions, diced

Mix the corn meal, flour, salt and pepper together. Dip the fish in this mixture, coating both sides well.

Heat half the oil in a skillet. Brown the onions in it. Remove onions and reserve. Add remaining oil to the skillet and brown the fish on both sides. Cover with the onions and cook 5 minutes. Serve hot or cold. Serves 6.

Baked Stuffed Brook Trout Meunière

¼ pound butter
½ cup minced onion
½ pound sliced mushrooms
1 tablespoon minced parsley
2 cups coarsely crushed soda crackers
3 teaspoons salt
½ teaspoon pepper
⅛ teaspoon thyme
4 brook trout
½ cup flour
2 tablespoons lemon juice

Melt half the butter in a skillet; cook the onion in it 5 minutes. Add the mushrooms and cook 5 minutes longer. Mix in the parsley, crackers, 1 teaspoon salt, ¼ teaspoon pepper and the thyme.

Split the trout for stuffing and stuff with the prepared mixture; sew the openings. Mix the flour with the remaining salt and pepper; roll the trout in it.

Melt the remaining butter in skillet; cook the trout in it over low heat, turning until browned on all sides. Bake 20 minutes in 350° oven. To serve, sprinkle with the lemon juice. Serves 4.

Deviled Halibut

¼ cup minced onion
3 tablespoons butter
2 tablespoons flour
1 cup milk
1½ teaspoons salt
1 teaspoon prepared mustard
2 teaspoons Worcestershire sauce
2 teaspoons lemon juice
1 egg, beaten
2 cups cooked flaked halibut
½ cup bread crumbs

Cook the onion in the butter for 5 minutes. Sprinkle with the flour and gradually add the milk, stirring constantly until mixture reaches the boiling point. Add the salt, mustard, Worcestershire sauce, lemon juice, egg and halibut. Mix lightly and divide among 6 buttered baking dishes. Sprinkle with the bread crumbs.

Bake in a 350° oven 20 minutes. Serves 6.

Mock Scallops

2 pounds halibut, ½ inch thick
¾ cup sifted flour
1½ teaspoons salt
¼ teaspoon pepper
¼ teaspoon paprika
1 cup milk
6 tablespoons butter

Cut the halibut in ½-inch cubes to resemble bay scallops.

Mix together the flour, salt, pepper and paprika. Dip the cubes in the milk and then in the seasoned flour.

Cook the scallops in the butter over low heat, until browned on all sides. Serve with tartar sauce and lemon wedges. Serves 6–8.

Baked Salmon

> 2 tablespoons butter
> 4 slices salmon
> 2 teaspoons salt
> ¼ teaspoon pepper
> 1 onion, sliced
> 1 bay leaf
> 1 cup light cream

Melt the butter in a baking dish. Arrange the salmon in it and sprinkle with the salt and pepper. Surround with the onion slices and bay leaf.

Bake in a 375° oven 15 minutes. Pour the cream over the fish and bake for additional 25 minutes, basting frequently. Serves 4–6.

Baked Fish and Vegetables

> ¼ cup salad oil
> 6 slices carp, pike or whitefish
> 3 teaspoons salt
> ¾ teaspoon pepper
> 2 tomatoes, diced
> 1 package frozen mixed vegetables, defrosted
> 3 potatoes, peeled and diced
> 2 onions, sliced
> ¼ teaspoon garlic powder

Heat the oil in a baking pan. Sprinkle the fish with half the salt and pepper and place in the pan. Bake in a 425° oven 10 minutes. Add the tomatoes, mixed vegetables, potatoes, onions, garlic powder and remaining salt and pepper. Cover. Reduce the heat to 350° and bake an additional 40 minutes, removing the cover for the last 15 minutes. Serves 6–8.

Sweet-and-Sour Fish

2 onions, thinly sliced
2 lemons, sliced
⅓ cup brown sugar
¼ cup seedless raisins
1 bay leaf
6 slices whitefish, pike or salmon
2 teaspoons salt
2 cups water
6 gingersnaps, crushed
⅓ cup cider vinegar
¼ cup sliced blanched almonds

Combine the onions, lemon slices, brown sugar, raisins, bay leaf, fish, salt and water in a saucepan. Cover and cook over low heat 25 minutes. Transfer fish to a platter.

Add the gingersnaps, vinegar and almonds to the fish stock. Cook over low heat, stirring constantly, until smooth. Pour over the fish. Serve warm or cold. Serves 6.

Pickled Pike

4 onions, sliced
6 slices pike
1½ teaspoons salt

¼ *teaspoon pepper*
2 *cups water*
½ *cup white vinegar*
1½ *tablespoons sugar*
2 *teaspoons pickling spice*
2 *bay leaves*

Combine 2 sliced onions, the pike, salt, pepper and water in a saucepan. Bring to a boil and cook over low heat 25 minutes.

Carefully remove the fish and place in a bowl or jar alternating in layers with the remaining 2 sliced onions.

Mix the vinegar, sugar, pickling spice and bay leaves with the fish stock. Bring to a boil and pour over the fish. cover and let pickle in the refrigerator for 2 days before serving. Serves 6 and will keep 2 weeks.

Marinated Salmon

6 *slices salmon*
3 *onions, sliced*
3 *cups water*
2 *teaspoons salt*
½ *cup lemon juice*
½ *cup white vinegar*
1 *teaspoon pickling spice*
1 *bay leaf*
¼ *teaspoon whole peppercorns*

Combine the salmon, onions, water and salt in a deep skillet. Bring to a boil and cook over low heat 25 minutes. Carefully transfer salmon to a bowl or platter.

Add the lemon juice, vinegar, pickling spice, bay leaf and peppercorns to the fish stock. Bring to a boil and cook 2 minutes. Pour over the fish and chill 24 hours before serving. Serves 6–8.

Fish Cakes

2 cups cooked and flaked codfish
2 cups mashed potatoes
3 tablespoons melted butter
2 teaspoons salt
½ teaspoon pepper
2 eggs, beaten
½ cup milk
Butter or fat for frying

Mix all the ingredients together and taste to correct seasoning. Shape into 8 cakes and chill for 2 hours.

Fry in the butter or fat until browned on both sides. Serve with hot or cold tomato sauce. Serves 4.

Seafood Newburg

4 tablespoons butter
2 tablespoons flour
2 cups light cream
1½ cups cubed cooked salmon
1½ cups cubed cooked halibut
1 teaspoon salt
⅛ teaspoon paprika
3 tablespoons sherry
3 egg yolks

Melt the butter in the top of a double boiler; stir in the flour, and gradually add 1 cup of the cream, stirring steadily until mixture reaches the boiling point. Add the salmon, halibut, salt, paprika and sherry. Place over hot water and cook 5 minutes.

Beat the egg yolks and remaining cream together; stir into the fish mixture and cook until just thickened. Serve on toast. Serves 6.

POULTRY

Chicken dishes are a staple with all Jewish families. In this section, you'll find many delicious but simple ways to prepare chicken in the Jewish style. Everyone knows that the classic Friday night meal of Jewish families throughout the world consists of chicken soup, followed by boiled or roast chicken. To serve anything else might almost border on the sacrilegious in the minds of many who have had no other Friday night dinner during their lifetimes.

Ducks, turkeys and geese are also important in Jewish cookery. Present methods of production have brought their prices down, and they offer a pleasant change for holiday fare, or for everyday meals, too.

Roast Pullet

> 2 cups mashed potatoes
> 2 tablespoons melted chicken fat
> ½ cup browned minced onions
> 3 teaspoons salt
> ½ teaspoon pepper
> 5-pound pullet
> ½ teaspoon ginger
> ½ cup boiling water

Mix together the potatoes, fat, onions, 1 teaspoon salt and ¼ teaspoon pepper. Stuff the pullet and sew the opening or fasten with skewers. Season with the remaining salt, pepper and the ginger. Place in a roasting pan.

Roast in a 425° oven 45 minutes, turning the pullet to brown on all sides. Add the water, reduce the heat to 350° and roast 1½ hours longer, or until pullet is tender. Baste occasionally. Serves 6.

Pot-Roasted Chicken

> 4-pound chicken
> 1 cup minced onions
> 4 tablespoons chicken fat
> 2 teaspoons salt
> ¼ teaspoon pepper
> 1 teaspoon paprika
> ¼ teaspoon garlic powder
> 2 teaspoons flour
> 2 cups boiling water

Have the chicken disjointed. Brown the onions and chicken in the fat. Sprinkle with the salt, pepper, paprika, garlic powder and flour. Stir in the water. Cover and cook over low heat 2 hours or until the chicken is tender. Serves 4–6.

Baked Chicken

> *3–4-pound fryer*
> *1 cup bread crumbs*
> *½ cup cracker meal*
> *2 teaspoons salt*
> *½ teaspoon pepper*
> *2 eggs, beaten*
> *⅓ cup chicken fat*
> *2 onions, thinly sliced*

Have the chicken disjointed. Mix the bread crumbs, cracker meal, salt and pepper together. Dip the chicken pieces in the eggs and then in the bread-crumb mixture.

Brown the chicken in the fat, removing the pieces as they brown. Lightly brown the onions in the fat remaining in the pan. Arrange the chicken in a baking dish with the onions around it.

Bake in a 350° oven 45 minutes. Serves 4.

Fried Chicken with Hot Cranberry Sauce

> *2 1½-pound broilers*
> *2½ teaspoons salt*
> *¼ teaspoon pepper*
> *1½ cups sifted flour*
> *1½ teaspoons baking powder*
> *1 egg*
> *½ cup water*
> *Fat for deep frying*
> *1 cup canned whole cranberries*
> *½ cup applesauce*

Have the broilers cut up. Season with 2 teaspoons salt, the pepper, and sprinkle with 3 tablespoons flour.

Sift the remaining flour, salt and the baking powder into a bowl. Beat in the egg and water. Dip the chicken pieces in this batter, coating them well.

Heat the fat to 370° and fry a few pieces at a time in it. Drain and place in a baking pan. Cover and bake in a 350° oven for 30 minutes, removing the cover for the last 10 minutes.

Combine the cranberries and applesauce and heat. Serve with the fried chicken. Serves 6–8.

Chicken Paprika

> 1 4-pound pullet
> 3 tablespoons flour
> 2 teaspoons salt
> ¼ teaspoon pepper
> 4 tablespoons chicken fat
> 1½ cups sliced onions
> 1 tablespoon paprika
> 1 cup boiling water

Have the chicken cut up and season with the flour, salt and pepper. Brown the chicken in the fat. Remove the chicken and brown the onions in the fat remaining in the pan. Return the chicken to the pan and sprinkle with the paprika and add the water. Cover and cook over low heat 1½ hours or until chicken is tender. Serve with spaetzel (pages 100–101). Serves 4–6.

Chicken à la King

> 4 tablespoons chicken fat
> 1 cup sliced mushrooms
> 4 tablespoons flour

½ *teaspoon salt*
⅛ *teaspoon pepper*
1¾ *cups chicken stock*
2 *cups diced cooked chicken*
1 *pimento, chopped*
2 *tablespoons sherry*

Heat the fat in a saucepan; cook the mushrooms in it for 5 minutes. Stir in the flour, salt and pepper. Gradually add the stock, stirring steadily until mixture reaches the boiling point. Cook over low heat 5 minutes. Add the chicken, pimento and sherry. Heat and serve on toast or in patty shells. Serves 4.

Chicken Chow Mein

2 *tablespoons oil*
1 *cup sliced onions*
1½ *cups sliced celery*
1 *cup sliced mushrooms*
2 *cups chicken stock*
3 *tablespoons soy sauce*
3 *cups cooked chicken, cut in strips*
1½ *tablespoons cornstarch*
½ *cup minced scallions*
Chow Mein noodles
Boiled white rice

Heat the oil in a saucepan; cook the onions, celery and mushrooms in it for 3 minutes. Add 1 cup of the stock, the soy sauce and chicken. Cook 5 minutes. Mix the cornstarch with the remaining stock and stir into the chicken mixture until thickened. Add the scallions. Arrange mixture over the noodles, which have been crisped in the oven. Serve with rice as a side dish. Serves 6.

Chicken Cantonese

4 tablespoons fat
4 tablespoons flour
3 cups chicken stock
3 cups diced chicken
1 cup water chestnuts, sliced
1 cup bamboo shoots, sliced
½ cup bean sprouts
½ cup sliced scallions
½ teaspoon pepper
Boiled white rice

Heat the fat and stir in the flour. Gradually add the stock, stirring constantly until mixture reaches the boiling point. Add the chicken, water chestnuts, bamboo shoots, bean sprouts, scallions and pepper. Cook over low heat 20 minutes. Taste to correct seasoning and serve on rice. Serves 6.

Fricassee of Giblets

2 pounds mixed giblets
4 tablespoons chicken fat
1 cup diced onions
2 tablespoons flour
4 cups boiling water
3 teaspoons salt
½ teaspoon pepper
½ teaspoon garlic powder
¾ pound ground beef
2 tablespoons cold water
¾ cup raw rice

Buy necks, gizzards, livers, wings and feet. Wash them thoroughly and remove the heavy skin from the feet.

Heat the chicken fat in a saucepan and brown the onions in it. Add the giblets and let them brown for 5 minutes. Sprinkle with the flour and add the boiling water, 2 teaspoons salt, ¼ teaspoon pepper and ¼ teaspoon garlic powder. Cover and cook over low heat 1 hour.

Mix the beef, cold water and remaining salt, pepper and garlic powder together. Shape into balls; add to the giblets with the rice. Cook 20 minutes. Serves 6.

Noodle-Stuffed Duck

> *6-pound duck*
> *3 teaspoons salt*
> *¾ teaspoon pepper*
> *1 teaspoon paprika*
> *½ teaspoon garlic powder*
> *¾ cup diced onions*
> *½ pound mushrooms, chopped*
> *4 tablespoons chicken fat*
> *2 tablespoons chopped parsley*
> *3 cups cooked medium noodles*
> *2 eggs, beaten*

Singe the duck to remove pin feathers; wash and dry. Sprinkle inside and out with 2 teaspoons salt, ½ teaspoon pepper, the paprika and garlic powder. Grind or chop the liver, gizzard and heart.

Cook the onions and mushrooms in the fat for 5 minutes. Add the ground giblets and cook 5 minutes longer. Remove from heat. Add the parsley, noodles, eggs and remaining salt and pepper. Mix lightly, then stuff the duck. Sew the opening or fasten with skewers. Place on a rack in a roasting pan.

Roast in a 375° oven 2½ hours, or until the duck is tender and brown. Turn duck twice during the roasting time. Serves 4.

Duck with Orange Sauce

> *1 5–6-pound duck*
> *2 teaspoons salt*
> *¼ teaspoon pepper*
> *3 tablespoons flour*
> *2 cups orange juice*
> *2 tablespoons lemon juice*
> *½ cup beef broth*
> *2 tablespoons currant jelly*
> *2 oranges, segmented*

Singe the duck to remove pin feathers; wash and dry. Season with the salt and pepper. Place on a rack in a roasting pan. Roast in a 350° oven 3 hours, or until tender and browned.

Measure 2 tablespoons of the fat from the roasting pan and pour it into a saucepan. Stir in the flour and gradually add the orange juice, mixing steadily until mixture reaches the boiling point. Add the lemon juice, broth, jelly and orange segments. Cook over low heat 10 minutes. Carve the duck and serve with the sauce. Serves 4.

Roast Duck with Apples

> *1 5–6-pound duck*
> *2 teaspoons salt*
> *¼ teaspoon pepper*
> *2 cloves garlic, minced*
> *3 cups diced apples*

12 pitted prunes
1 egg
4 tablespoons dry bread crumbs
2 tablespoons brown sugar

Singe the duck to remove pin feathers; wash and dry. Make a paste of the salt, pepper and minced garlic. Rub into the duck. Mix together the apples, prunes, egg, bread crumbs and sugar. Stuff the duck and close the opening. Place on a rack in a roasting pan.

Roast in a 400° oven 20 minutes; reduce the temperature to 350° and roast 2½ hours longer or until browned and tender. Pour off the fat a few times during roasting. Serves 4.

Roast Turkey

12-pound turkey
1 tablespoon salt
½ teaspoon pepper
1 pound unsweetened pitted prunes
1 cup water
4 cups sliced apples
1 cup dry bread crumbs
2 teaspoons lemon juice
1 tablespoon sugar
1 teaspoon cinnamon

Season the turkey with the salt and pepper. Cook the prunes in the water for 10 minutes. Drain. Add the apples, bread crumbs, lemon juice, sugar and cinnamon. Mix lightly and stuff the turkey. Sew the opening or fasten with skewers. Place in a roasting pan.

Roast in a 350° oven 3–4 hours or until the turkey is tender. Turn turkey to brown on all sides. Serves 10–12.

Goose Fricassee

10-pound goose
½ cup flour
2 teaspoons salt
½ teaspoon pepper
1 teaspoon paprika
½ teaspoon garlic powder
3 tablespoons rendered goose fat
1 cup diced onions
2 cups boiling water
1 small bay leaf

Buy a young goose and have it disjointed. Remove all the skin. (You can render it with the fat.) Mix the flour, salt, pepper, paprika and garlic powder together. Lightly roll the pieces of goose in the mixture. Brown the goose and onions in 3 tablespoons of the fat. Add the water and bay leaf. Cover and cook over low heat 2½ hours, or until the goose is tender. Remove bay leaf. Serves 8–10.

Roast Stuffed Goose

12-pound goose
4 tablespoons rendered goose fat
1 cup diced onions
5 cups grated drained potatoes
1 egg, beaten
4 teaspoons salt
1 teaspoon pepper
1 tablespoon paprika
½ teaspoon garlic powder

Remove as much fat from the goose as possible. (See page 58 on how to render fat.) Grind or chop the liver and gizzard. Heat 4 tablespoons of the fat in a skillet and lightly brown the onions in it. Stir in the potatoes and cook over low heat 2 minutes, stirring constantly. Cool 5 minutes, then add the egg, 1½ teaspoons salt, ¼ teaspoon pepper and 1 teaspoon paprika.

Sprinkle the garlic powder and the remaining salt, pepper and paprika on the goose, then stuff it. Sew the openings or fasten with skewers. Place on a rack in a roasting pan. Cover the pan and roast in a 350° oven 30 minutes. Remove the cover and roast 2½ hours longer or until the goose is tender and brown. Serves 8–10.

Goose with Cabbage

8–10-pound goose
3 teaspoons salt
½ teaspoon pepper
1 cup sliced onions
6 pounds cabbage, shredded
1 tart apple, grated

Have the goose cut up and season it with 2 teaspoons salt and ¼ teaspoon pepper. Place on a rack in roasting pan and roast in a 350° oven 2 hours or until almost tender.

Measure 6 tablespoons of the fat from the roasting pan into a heavy casserole or Dutch oven and cook the onions and cabbage in it over low heat for 20 minutes. Add the apple, remaining salt and pepper. Arrange the goose on top. Cover and cook over low heat 1 hour. Serves 8–10.

Cornish Hen with Wild Rice

> 1 cup wild rice
> ¼ cup minced onion
> ¼ cup chopped mushrooms
> ground giblets of hen
> 4 tablespoons chicken fat
> 3 teaspoons salt
> ¾ teaspoon pepper
> 1 3-pound Cornish hen or 2 small Cornish hens
> ½ cup boiling water

Cook the wild rice according to directions on package. Drain.

Cook the onion, mushrooms and giblets in 2 tablespoons chicken fat for 10 minutes. Stir in the rice, 1 teaspoon salt and ¼ teaspoon pepper.

Season the hen or hens with the remaining salt and pepper. Stuff and sew the opening.

Heat the remaining fat in a deep skillet and brown the hen in it. Add ½ cup boiling water. Cover. Bake in a 400° oven 45 minutes or until tender. Remove cover for last 15 minutes. Serves 3–4.

Rendered Fat and Grebenes

Remove fat and the fatty skin from chicken or goose. For each cup of fat to be rendered, you'll need ¼ cup of sliced onions and 1 slice apple. Wash and drain the fat and cut in small pieces. Cook over low heat until the fat is almost melted. Add the onions and apple and cook until the onions brown. Cool and strain. The onions and pieces of skin (grebenes—cracklings) can be stored in refrigerator for use in a number of dishes. They are delicious in such dishes as Kasha Varnitchkes, Flaishig Noodle Pudding, Chopped Liver, etc.

Stuffed Helzel or Derma

Carefully remove the neck skin of a goose or two or three chickens. Sew together one end. Stuff with the following mixture, then sew the other end. Roast in the pan with the goose. Cut in slices to serve.

> 1½ *cups sifted flour*
> 4 *tablespoons grated onion*
> ½ *cup rendered goose fat*
> 1¼ *teaspoons salt*
> ¼ *teaspoon pepper*
> 1 *teaspoon paprika*

Lightly mix all the ingredients together.

MEAT

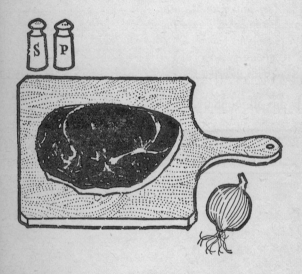

ınterestingly seasoned meat dishes rank high among the favorites in the Jewish cuisine. Broiled steaks and rib roasts of beef are duly appreciated, but those are American additions and not indigenous to the basic traditions of Jewish cooking. In most families, the choice would be a meat dish with a seasoned gravy and perhaps some dumplings to satisfy the appetite. And how good these dishes taste the next day either in a sandwich or reheated in their own gravy!

Boiled Flanken

    4 pounds flank
    2 tablespoons fat
    1½ quarts boiling water
    2 onions
    2 carrots
    3 stalks celery
    3 sprigs parsley
    1 tablespoon salt
    6 whole peppercorns

Lightly brown the flank in the fat. Drain. Add the water,
onions, carrots, celery, parsley, salt and peppercorns.
Cover and cook over low heat 2 hours or until the meat is
tender. Strain the stock and serve as soup or use in other
dishes. Serve the meat with horseradish. Serves 6–8.

Gedempte Flaish

    4 pounds chuck or brisket
    2 tablespoons fat
    3 onions, diced
    2 teaspoons salt
    ¼ teaspoon pepper
    1 teaspoon paprika
    ½ teaspoon garlic powder
    1 bay leaf
    1½ cups boiling water
    1 tablespoon browned flour*

* Note: Spread one cup of sifted flour in a shallow pan; cook over
low heat, stirring almost constantly. Cool and store in a covered
jar. Use for thickening gravies—it gives a much nicer color than
plain flour.

Place the meat and fat in a heavy saucepan or Dutch oven. Cover and brown over medium heat until browned on all sides. Add the onions and continue browning. Add the salt, pepper, paprika, garlic powder, bay leaf and water. Cover and cook over low heat 2 hours or until meat is tender.

Mix the flour with a little water and stir into the gravy. Cook 10 minutes. Remove bay leaf. Serves 8.

Gedempte Flaish with Apricots

1 pound dried apricots
4 cups water
2 onions, minced
3 pounds center-cut chuck
2 tablespoons fat
1½ teaspoons salt
1 bay leaf
1 tablespoon lemon juice
2 tablespoons brown sugar
1 teaspoon cinnamon

Wash the apricots and soak in the water for 1 hour.

Brown the onions and meat in the fat. Add the salt, bay leaf, lemon juice, sugar, cinnamon and undrained apricots. Cover and cook over low heat 2½ hours or until meat is tender. Remove bay leaf. Serves 6–8.

Essig Flaish

4 pounds chuck, brisket or flank
6 onions, diced
1½ teaspoons salt

3 cups boiling water
⅓ cup lemon juice
3 tablespoons brown sugar
4 gingersnaps, crushed

Heat a heavy saucepan or Dutch oven and brown the meat in it over medium heat. If meat is very lean add a little fat for browning. Turn frequently. Add the onions and brown lightly. Add the salt and water. Cover and cook over low heat 2 hours. Stir in the lemon juice, brown sugar and gingersnaps. Cook 10 minutes. Taste to correct seasoning. Serves 8.

Pot Roast with Vegetables

2 teaspoons salt
½ teaspoon pepper
½ teaspoon garlic powder
4 pounds chuck or brisket
2 tablespoons fat
3 onions, sliced
1½ cups boiling water
3 tomatoes, diced
3 carrots, quartered
2 green peppers
4 potatoes, peeled and cut in half

Sprinkle the salt, pepper and garlic powder on the meat. Heat the fat in a heavy saucepan or Dutch oven and brown the meat in it. Add the onions and continue to brown. Add the water and tomatoes. Cover and cook over low heat 2½ hours. Add the carrots, green peppers and potatoes. Cook 30 minutes. Taste to correct seasoning. Serves 8–10.

Beef Goulash

> 3 pounds chuck
> 2 tablespoons fat
> 5 onions, diced
> 1½ teaspoons salt
> 2 teaspoons paprika
> ¼ teaspoon pepper
> 2 green peppers, sliced
> 1 cup canned tomato sauce

Cut the beef in 2-inch cubes. Heat the fat in a heavy saucepan or Dutch oven and brown the meat in it. Stir in the onions and continue to brown. Sprinkle with the salt, paprika and pepper. Cover and cook over low heat ½ hour. Add the green peppers and tomato sauce. Cook 1½ hours longer or until meat is tender. Serve with noodles or boiled potatoes. Serves 6–8.

Sauerkraut Beef

> 3 pounds chuck
> 3 onions
> 2 tablespoons fat
> 1½ teaspoons salt
> ¼ teaspoon pepper
> 1 teaspoon paprika
> 1½ pounds sauerkraut
> 1 bay leaf
> 1 cup boiling water

Cut the beef in 2-inch cubes and brown with the onions in the fat. Sprinkle with the salt, pepper and paprika. Cover and cook over low heat 30 minutes. Stir in the sauerkraut. Cook 10 minutes. Add the bay leaf and water. Cover and cook an additional 1½ hours. Remove bay leaf. Serves 6–8.

Lentil and Rice Stew

> 3 pounds chuck
> 3 onions, diced
> 2 tablespoons fat
> 1 cup lentils
> 2 teaspoons salt
> ½ teaspoon pepper
> 4 cups boiling water
> ½ cup uncooked rice

Cut the beef in 2-inch cubes and brown with the onions in the fat. Soak lentils in cold water for 2 hours. Drain. Add the lentils, salt, pepper and boiling water. Cover and cook over low heat 1½ hours. Stir in the rice; recover and cook additional 25 minutes. Serves 6–8.

Pepper Steak

> 2 pounds steak
> 4 tablespoons oil
> ½ cup scallions or onions
> 2 cloves garlic, minced
> 5 green peppers, thinly sliced
> 1 cup sliced celery
> 1½ cups beef broth
> 2 tablespoons cornstarch
> ¼ cup water
> 2 tablespoons soy sauce
> Boiled white rice

Buy thin steak and cut it in narrow strips. Brown the steak in the oil, then add the scallions, garlic, green peppers and celery. Cook 5 minutes. Add the beef broth. Cover and cook over low heat 10 minutes. Mix the cornstarch, water and soy sauce until smooth and add to the mixture, stirring steadily until it reaches the boiling point. Cook 2 minutes and serve on rice. Serves 6–8.

Home-Style Hamburgers

> 1½ pounds ground beef
> 4 tablespoons grated onion
> 2 cloves garlic, minced
> 2 teaspoons salt
> ¼ teaspoon pepper
> ¼ cup grated potato, drained
> 1 egg
> ¼ cup water
> 3 tablespoons fat
> 2 onions, sliced

Mix the beef, onion, garlic, salt, pepper, potato, egg and water together. Form into 8 hamburgers.

Melt the fat in a skillet and brown the hamburgers and onion for 15 minutes, turning the hamburgers after 10 minutes. Serves 4–8.

Sweet-and-Sour Meat Balls

> 1½ pounds ground beef
> 4 tablespoons grated onion
> 1 egg
> 2 teaspoons salt

⅛ teaspoon pepper
3 tablespoons cornstarch
2 tablespoons fat
1½ cups beef stock
2 tablespoons lemon juice
1 lemon, thinly sliced
¼ cup seedless raisins
3 tablespoons sugar
6 gingersnaps, crushed

Mix the meat, onion, egg, 1 teaspoon salt and the pepper. Form mixture into 1-inch balls; lightly roll them in the cornstarch.

Melt the fat in a deep skillet. Brown the meat balls in it. Add the stock, lemon juice, lemon, raisins and sugar. Cover and cook over low heat 35 minutes. Stir in the gingersnaps and cook additional 10 minutes. Serves 6.

Stuffed Cabbage

Large head cabbage
2 tablespoons fat
2 onions, sliced
3 cups canned tomatoes
3 teaspoons salt
½ teaspoon pepper
Beef bones
1 pound ground beef
3 tablespoons uncooked rice
4 tablespoons grated onion
1 egg
3 tablespoons cold water
3 tablespoons honey
¼ cup lemon juice
¼ cup seedless raisins

Pour boiling water over the cabbage to cover and let soak for 15 minutes. Remove 12 leaves carefully; if leaves are small, use 18.

Heat the fat in a deep, heavy saucepan. Lightly brown the onions in it. Add the tomatoes, half of the salt and pepper and all the bones. Cook over low heat 30 minutes.

Mix together the beef, rice, grated onion, egg and water.

Place some of the meat mixture on each cabbage leaf. Tuck in the sides and roll up carefully. Add to the sauce. Cover and cook over low heat 1½ hours. Add the honey, lemon juice and raisins. Cook 30 minutes longer.

Serves 6 as a main course, 12 as a first course.

Stuffed Peppers

Substitute 12 peppers for the cabbage. Cut a 1-inch piece from the stem end and reserve. Scoop out the seeds and fibers carefully. Make the sauce as directed in the Stuffed Cabbage recipe. Stuff the peppers with the meat mixture, top with stem ends and proceed as directed.

Serves 6 as a main course, 12 as a first course.

Stuffed Breast of Veal

> 5 pounds breast of veal
> 2½ teaspoons salt
> ½ teaspoon pepper
> 1 teaspoon paprika
> ½ teaspoon garlic powder
> 3 tablespoons fat

Have the butcher make a pocket in the veal. Sprinkle with the salt, pepper, paprika and garlic powder. Fill the pocket with one of the following stuffings and fasten the opening with skewers or thread.

Melt the fat in a roasting pan and place the veal in it. Roast in a 325° oven 3 hours or until meat is tender. Baste frequently and add a little water if pan becomes dry. Serves 6–8.

POTATO STUFFING:

>   2 cups grated potato, drained
>   ½ cup minced onion
>   4 tablespoons potato flour
>   1 egg
>   1½ teaspoons salt
>   ¼ teaspoon pepper

Mix all the ingredients and stuff the veal.

BREAD STUFFING:

>   1 onion, minced
>   1 green pepper, diced
>   3 stalks celery, sliced
>   3 tablespoons fat
>   6 slices bread
>   1 teaspoon salt
>   ⅛ teaspoon pepper
>   ⅛ teaspoon thyme
>   1 teaspoon paprika
>   1 egg

Cook the onion, green pepper and celery in the fat for 10 minutes, stirring occasionally. Soak the bread in water; squeeze dry and pull into small pieces. Add to the vegetables with the salt, pepper, thyme, paprika and egg. Mix well and stuff the veal.

Veal Burgers

> ¾ cup minced onions
> 6 tablespoons fat
> 4 slices white bread
> 1½ pounds ground veal
> 1½ teaspoons salt
> ¼ teaspoon pepper
> 3 eggs, beaten
> ¾ cup matzo or cracker meal

Brown the onions in half the fat. Wet the bread and mash until smooth. Mix the browned onions, bread, veal, salt, pepper and 2 eggs. Shape into hamburgers. Dip in the remaining beaten egg and then in the meal.

Heat the remaining fat in a skillet. Cook over low heat until browned on both sides. Serves 6.

Veal Goulash

> 2 pounds veal
> 2 onions, diced
> 3 tablespoons fat
> 1½ teaspoons salt
> ¼ teaspoon pepper
> 2 teaspoons paprika
> 2 green peppers, sliced
> ¾ cup canned tomato sauce

Cut the veal in 1½-inch cubes. Brown the veal and onions in the fat. Add the salt, pepper, paprika, green peppers and tomato sauce. Cover and cook over low heat 1 hour or until veal is tender. Serve with noodles or dumplings. Serves 4–6.

Baked Veal Chops

1 cup matzo or cracker meal
2 teaspoons salt
¼ teaspoon pepper
2 eggs
4 veal chops
4 tablespoons fat
2 onions, thinly sliced

Mix the meal, 1½ teaspoons salt and the pepper together. Beat the eggs and remaining salt. Dip the chops in the meal, the eggs and then the meal again.

Heat the fat in a baking pan. Arrange the chops in it and surround with the sliced onions. Bake in a 350° oven 45 minutes, turning the chops to brown evenly. Serves 4.

Gedempte Veal

3 pounds veal
4 onions, thinly sliced
3 tablespoons fat
1½ teaspoons salt
¼ teaspoon pepper
2 teaspoons paprika
¾ cup water

Cut the veal in 1-inch cubes. Brown the veal and onions in the fat. Add the salt, pepper, paprika and water. Cover and cook over low heat 1 hour or until the veal is tender. Stir frequently. Serve with boiled potatoes or farfel. Serves 6–8.

Lamb Stew

3 pounds boneless lamb
¼ cup flour
2 teaspoons salt
½ teaspoon pepper
1 teaspoon paprika
½ teaspoon garlic powder
3 tablespoons fat
1 cup diced onions
1 cup canned tomato sauce
1 cup boiling water
1 bay leaf
1 cup sliced green peppers
3 potatoes, peeled and sliced
1 package frozen mixed vegetables

Cut the lamb into 2-inch cubes. Mix together the flour, salt, pepper, paprika and garlic powder. Lightly roll the lamb in this mixture, then brown in the fat with the onions. Add the tomato sauce, water, bay leaf and green peppers. Cover and cook over low heat 1½ hours. Add the potatoes and vegetables. Cook 20 minutes. Remove bay leaf. Serves 6–8.

Pickled Tongue or Corned Beef

6 pounds brisket or fresh tongue
1¼ cups salt
2 tablespoons pickling spice
1 teaspoon saltpeter
1 teaspoon sugar
12 cloves garlic
1 quart water

Place the meat in a large stone crock. Mix the salt, pickling spice, saltpeter, sugar and garlic with 1 quart water and pour over the meat. Add enough water to completely cover the meat. Use a heavy bowl or board to weight the meat down. Cover the crock with cheesecloth, tie it in place, then cover the cheesecloth with aluminum foil.

Let stand in a cool place for 8 days, then store in refrigerator for 6 days.

To serve, cook in boiling water for 3 hours or until tender. Serves 8–10.

Corned Beef

1½ cups salt
4 quarts water
1 tablespoon sugar
2 tablespoons pickling spice
½ ounce saltpeter
8 bay leaves
5 pounds first cut brisket of beef
8 cloves garlic
2 onions
2 stalks celery

Combine the salt, water, sugar, pickling spice, saltpeter and bay leaves in a saucepan. Bring to a boil and cook for 5 minutes. Cool. Place the beef in a stone crock or bowl (not metal). Pour the cool liquid over it and add the garlic. Weight the meat down to keep it covered by the liquid. Cover with a piece of muslin and tie. Let pickle for 12 days in cool place, preferably a refrigerator.

TO COOK:

Rinse the meat, add the onions and celery, cover with water. Bring to boil and cook on low heat for 3 hours or until tender. Drain and slice crosswise. Cabbage and potatoes can be cooked in the stock and served with the corned beef. If you use the entire five pounds, it will serve 12–14.

Some people think cold corned beef is even better than hot, and others think it's best of all in a sandwich, made, of course, with tangy rye bread,* hot English-style mustard, and a pickle or some cold sauerkraut on the side. Another favorite sandwich is pastrami, but for this you'd better get your pastrami at your favorite delicatessen, since it's too difficult to prepare at home.

Boiled Pickled Tongue

> *4–5-pound pickled tongue*
> *1 onion*
> *2 cloves garlic*
> *2 bay leaves*

Wash the tongue. Combine with onion, garlic and bay leaves in a deep kettle. Cover with water. Bring to a boil and cook over medium heat 3½ hours or until tender. Add boiling water as it cooks out.

Let the tongue cool in the stock, then remove the root and skin. Reserve 2 cups stock if you want to make a sauce. Remove bay leaves. Slice the tongue and serve plain or with a sauce. Serves 6–8.

---

* We make a rye bread of this kind that we think is pretty good. It's called Grossinger's Country Club Rye. Try it, if it is available where you live.

Roast Tongue

> 4–5-pound fresh tongue
> 3 onions, sliced
> 2 teaspoons salt
> ½ teaspoon pepper
> ¾ teaspoon garlic powder
> 1 bay leaf
> 1 cup canned tomato sauce
> 2 tomatoes, diced
> 2 cups boiling water

Plunge the tongue into boiling water and cook 10 minutes. Drain and remove the skin and root.

Place the tongue in a roasting pan with the onions around it. Sprinkle with the salt, pepper and garlic powder. Add the bay leaf, tomato sauce, tomatoes and water. Cover the pan and roast in a 325° oven 3 hours or until the tongue is tender. Baste frequently and remove the cover for the last 30 minutes. Discard bay leaf. Serves 6–8.

Tongue with Sweet-and-Sour Sauce

> 2 tablespoons fat
> 1 onion, diced
> 2 tablespoons flour
> 2 cups tongue stock (page 74)
> ⅓ cup vinegar
> ⅓ cup honey
> ½ teaspoon salt
> ½ teaspoon powdered ginger

¼ cup seedless raisins
¼ cup sliced blanched almonds
1 lemon, thinly sliced
Cooked tongue

Melt the fat in a saucepan and lightly brown the onion. Sprinkle the flour on the browned onion; gradually add the stock, stirring constantly until mixture boils. Stir in the vinegar, honey, salt, ginger and raisins. Cook over low heat 5 minutes. Add the almonds and lemon. Cook 2 minutes.

Slice the tongue and serve with the sauce. Serves 6.

Sweet-and-Sour Calf's Liver with Spaetzel

2 onions, sliced
3 tablespoons fat
1 pound calf's liver, diced
1½ teaspoons salt
¼ teaspoon pepper
¼ teaspoon paprika
2 tablespoons flour
1½ cups boiling water
2 tablespoons lemon juice
2 teaspoons sugar

Lightly brown the onions in the fat. Add the liver and cook over medium heat 5 minutes. Sprinkle with the salt, pepper, paprika and flour. Add the water, lemon juice and sugar, stirring constantly until mixture reaches the boiling point. Cook over low heat 5 minutes. Taste to correct seasoning and serve with spaetzel (pages 100–101). Serves 4.

Calf's Liver with Onion Sauce

1 pound calf's liver
½ cup flour
1½ teaspoons salt
¼ teaspoon pepper
4 tablespoons chicken fat
3 cups sliced onions
1 tablespoon cornstarch
1 cup beef broth

Have the liver sliced into 4 thin slices. Dip in the flour mixed with the salt and pepper.

Heat the fat in a skillet and brown the liver on both sides. Remove liver. Brown the onions in the fat remaining in the pan. Mix the cornstarch and broth together and add to the onions, stirring constantly until mixture reaches the boiling point. Scrape the bottom of skillet and cook additional 2 minutes. Pour over the liver. Serves 4.

Sunday Sweetbreads

4 pairs calf's sweetbreads
1 tablespoon vinegar
3 cups water
3 teaspoons salt
4 tablespoons chicken fat
1 cup diced onions
1 pound mushrooms, sliced
2 tablespoons potato flour
¼ teaspoon pepper
½ teaspoon paprika
3 tablespoons minced parsley

Wash the sweetbreads and soak in cold water for 20 minutes. Drain.

Combine the sweetbreads, vinegar, 3 cups water and 2 teaspoons salt in a saucepan. Bring to a boil and cook over low heat 20 minutes. Remove sweetbreads and plunge into ice water for 20 minutes. Measure 2 cups of the stock and reserve. Remove the membrane and dice the sweetbreads.

Melt 2 tablespoons fat in a skillet and brown the onions in it. Remove the onions and reserve. Melt the remaining fat in the skillet. Cook the mushrooms in it for 5 minutes. Mix the potato flour and stock until smooth and stir into the mushrooms. Cook over low heat 5 minutes, stirring constantly. Add the browned onions, the sweetbreads, remaining salt, the pepper, paprika and parsley. Heat and serve on toast or in patty shells. Serves 8–10 as a first course.

Kishke

> 3 feet of beef casing
> 1 cup sifted flour
> ½ cup matzo or cracker meal
> ¼ cup grated onion
> 1½ teaspoons salt
> ¼ teaspoon pepper
> 1 teaspoon paprika
> 1 cup chicken fat
> 2 onions, sliced

Wash the casing in cold water and scrape the inside. Cut casing in half and sew one end of each half.

Blend well the flour, meal, grated onion, salt, pepper, paprika and ¾ cup of fat. Stuff the casings and sew the open ends. Cook in boiling salted water 1 hour. Drain.

Spread the remaining fat and the sliced onions in a

baking dish. Arrange the kishke over it. Roast in a 350° oven 1½ hours, basting frequently. Or, if you prefer, you can roast it in the same pan with meat or poultry with which it will be served. Slice and serve. Serves 8–10.

Barbecued Ribs of Beef

3 pounds short ribs
2 teaspoons salt
¼ teaspoon pepper
1 teaspoon paprika
1 teaspoon dry mustard
1 tablespoon sugar
1 tablespoon Worcestershire sauce
½ cup catchup
½ cup water
¼ cup cider vinegar
½ cup minced onions
1 clove garlic, minced

You can use the top of the roast beef bones. Have them cut in 2-inch pieces.

Brown the ribs in a heated casserole or Dutch oven; pour off the fat. Combine all the remaining ingredients and add to ribs. Cover and bake in a 350° oven 2 hours. Remove the cover for the last half hour. Serves 6–8.

Salami and Eggs

4 slices salami
2 eggs
⅛ teaspoon salt
Dash pepper
Fat for frying

Cut the salami in strips. Lightly beat the eggs, salt and pepper.

Fry the salami until lightly browned in a skillet. Pour the eggs over it and cook over low heat until the eggs are set on the bottom. Turn over, pancake fashion, and cook until set. Serves 1.

Frankfurter Hash

>   12 frankfurters
>   4 boiled potatoes, peeled and diced
>   3 onions, sliced
>   2 tablespoons fat
>   3 green peppers, sliced
>   3 tomatoes, diced
>   1 teaspoon salt
>   ¼ teaspoon pepper

Cut the frankfurters into 1-inch slices.

Lightly brown the potatoes and onions in the fat. Add the green peppers and frankfurters. Cook over low heat 5 minutes. Add the tomatoes, salt and pepper. Cover and cook 10 minutes. Serves 6.

Chile con Carne

>   4 tablespoons salad oil
>   1 cup thinly sliced onions
>   ½ cup diced green pepper
>   1 pound ground chuck
>   1½ cups barley water
>   2 cups canned tomatoes
>   3 tablespoons chile powder

1 *teaspoon salt*
2 *teaspoons sugar*
2 *cloves garlic, minced*
4 *cups cooked or canned kidney beans*

Heat the oil in a heavy casserole or Dutch oven; cook the onions and green pepper for 10 minutes. Add the meat and cook over high heat, stirring almost constantly until browned. Add the water, tomatoes, chile powder, salt, sugar and garlic. Cover and cook over low heat 1 hour. Add the beans and cook 30 minutes. Serves 8–10.

CHOLENT

Religious Jewish people are not permitted to cook on the Sabbath. However, dishes prepared in advance may be kept hot in a previously lit oven. In Central Europe, one of the favorite Sabbath dishes was *cholent* (primarily a bean dish) because its flavor was not impaired by long, slow cooking; if anything, it was improved. The good housewives would prepare their *cholent* on Friday afternoon and place it in the local baker's ovens; the fires were banked but the ovens retained their intense heat over the Sabbath. It would cook slowly overnight and after *schul* (synagogue) services were finished, it would be a delicious hot dish for a hungry family.

Cholent may be served as a main course or as a side dish, particularly with roast meats. Its consistency when done is quite thick, without liquid, but not quite dry. In a general sort of way, it might be said to resemble old-fashioned Boston baked beans, although it isn't quite so sweet. The following *cholent* recipes include meat, a modern refinement on a dish once composed exclusively of beans simply because many Jewish families could not afford the meat.

Meat Cholent

> 2 *cups dried lima beans*
> 3 *pounds brisket*
> 3 *onions, diced*
> 3 *tablespoons fat*
> 2 *teaspoons salt*
> ¼ *teaspoon pepper*
> ¼ *teaspoon ginger*
> 1 *cup pearl barley*
> 2 *tablespoons flour*
> 2 *teaspoons paprika*

Soak the beans overnight in water to cover. Drain.

Use a heavy saucepan or Dutch oven and brown the meat and onions in the fat. Sprinkle with the salt, pepper and ginger. Add the beans and barley and sprinkle with the flour and paprika. Add enough boiling water to cover one inch above the mixture. Cover tightly.

Cholent may be baked for 24 hours in a 250° oven or, for quicker cooking, bake in a 350° oven 4–5 hours. Slice the meat and serve with the barley and beans. Serves 8–10.

Potato Cholent

> 6 eggs
> 1 cup melted fat
> 2 cups sifted flour
> 3 teaspoons salt
> 3 pounds flanken
> 6 potatoes, peeled and cut in half
> ½ teaspoon pepper
> 2 teaspoons paprika
> ½ teaspoon garlic powder

Beat the eggs, fat, flour and 1 teaspoon salt together. Form into a flat mound and place in the center of a Dutch oven or baking dish. Place the meat on one side of the dough and arrange the potatoes around it. Sprinkle with the pepper, paprika, garlic powder and remaining salt. Add enough boiling water to cover all the ingredients. Cover tightly and bake in a 250° over 24 hours or in a 350° oven 4–5 hours.

Slice the meat, cut the crust and serve with the potatoes. Serves 6–8.

Cholent with Knaidel

> 2 cups lima beans
> 3 pounds flanken
> 3 onions, sliced
> ¾ cup fat
> 3 potatoes, peeled and halved
> 3 teaspoons salt
> ½ teaspoon pepper
> 2 eggs

  2 teaspoons sugar
  ¼ cup water
  ¾ cup matzo or cracker meal

Soak the beans overnight in water to cover. Drain.

Use a Dutch oven or baking dish and brown the meat and onions in ¼ cup fat. Add the potatoes and 2 teaspoons salt and the pepper.

Beat the eggs, sugar, remaining fat and remaining salt together. Stir in the water and meal. Form into a ball and place in the baking dish with other ingredients. Add boiling water to cover. Cover tightly and bake in a 250° oven 24 hours or in a 350° oven 4–5 hours. Serves 6–8.

Quick Kasha Cholent

  1 cup dried lima beans
  4 pounds chuck or brisket
  2 onions, sliced
  2 tablespoons fat
  1 tablespoon salt
  ½ teaspoon pepper
  1 teaspoon paprika
  ½ teaspoon garlic powder
  5 cups boiling water
  1 cup kasha

Soak the beans in water to cover for 12 hours. Drain.

Brown the meat and onions in the fat. Add the salt, pepper, paprika, garlic powder, water and beans. Cover loosely and cook over low heat 2 hours. Add the kasha and a little more water if necessary.

Cook ½ hour longer or until meat and beans are tender. Slice the meat and serve with the beans and kasha. Serves 8–10.

Quick Lamb Cholent

>   2 cups dried lima beans
>   2 lamb shanks
>   2 onions, sliced
>   3 tablespoons fat
>   2 cloves garlic, minced
>   2 teaspoons salt
>   ¼ teaspoon pepper
>   ½ teaspoon ginger
>   3 cups water

Soak the beans overnight in water to cover. Bring to a boil and cook 30 minutes. Drain.

Brown the lamb and onions in the fat. Add the garlic, salt, pepper, ginger and water. Cover and cook over low heat 2½ hours or until lamb and beans are tender. Add a little water if necessary. Serves 6–8.

DAIRY DISHES

Eggplant Steak à la Meyer

> 1 medium-sized eggplant
> 1 teaspoon salt
> ¼ teaspoon pepper
> 2 egg yolks, beaten
> 1 cup grated American cheese

Peel the eggplant and cut lengthwise in 1-inch slices.
Soak in cold salted water for 1 hour. Drain and dry.

Place the eggplant slices side by side in a buttered
baking pan. Sprinkle with the salt and pepper and brush
with the egg yolks. Sprinkle with the grated cheese. Bake
in a 350° oven 30 minutes. Top with a fried egg for each
serving. Serves 4–6.

Mushrooms in Sour Cream

>   ¾ cup diced onions
>   4 tablespoons butter
>   1½ pounds mushrooms, sliced
>   ¾ teaspoon salt
>   ¼ teaspoon pepper
>   1 teaspoon paprika
>   1 cup sour cream

Cook the onions in half the butter for 10 minutes. Remove the onions and reserve. Add the remaining butter and the mushrooms to the same pan and cook over low heat 10 minutes. Return the onions to the pan and add the salt, pepper and paprika. Cook 5 minutes. Stir in the sour cream. Serve on toast or in patty shells. Serves 3–4.

Vegetable Cutlet

>   1 cup chopped onion
>   ½ cup chopped celery
>   1 cup grated carrots
>   2 tablespoons butter
>   ½ cup cooked green beans, coarsely chopped
>   ½ cup cooked green peas
>   3 eggs
>   2 teaspoons salt
>   ½ teaspoon pepper
>   4 tablespoons matzo meal
>   Fat for frying

Cook the onion, celery and carrots in the butter for 10 minutes. Add the beans, peas, 2 beaten eggs, the salt, pepper and matzo meal. Mis well and shape into 6 cutlets. Beat the remaining egg and carefully dip the cutlets in it. Fry in hot fat until browned on both sides. Serve with your favorite mushroom sauce. Serves 6.

Rice-Cheese Casserole

> 3 cups grated carrots
> 2 cups cooked rice
> 2 eggs, beaten
> ½ cup light cream
> 1½ teaspoons salt
> ¼ teaspoon pepper
> 3 tablespoons grated onion
> 2 cups grated American cheese

Combine the carrots, rice, eggs, cream, salt, pepper, onion and 1½ cups cheese. Pour into a greased 1½-quart casserole. Sprinkle with the remaining cheese.

Bake in a 350° oven 50 minutes. Serves 6.

Spanish Rice

> ¼ cup olive oil
> ¾ cup rice
> 1 cup thinly sliced onions
> ½ cup chopped green pepper
> 1 can (#2½) tomatoes
> 1½ cups water
> 1½ teaspoons salt
> ¼ teaspoon pepper
> 1 bay leaf

Heat the oil in a casserole. Add the rice, onions and green pepper, stirring constantly, until browned. Add the tomatoes, water, salt, pepper and bay leaf. Cover.

Bake in a 350° oven 1 hour, removing the cover for the last 20 minutes. Remove bay leaf. Serves 4–6.

Spaghetti with Vegetarian Meat Balls

3 diced onions
½ pound mushrooms, sliced
4 tablespoons olive oil
1 can (#2½) tomatoes
1 can tomato paste
½ teaspoon oregano
2 onions, chopped
1 stalk celery, chopped
3 carrots, grated
6 tablespoons butter
3 eggs, beaten
1½ cups matzo meal
2 cups cooked green peas
1 teaspoon salt
¼ teaspoon pepper
1 pound spaghetti, cooked and drained
Grated Swiss cheese

Cook the diced onions and mushrooms in the oil for 10 minutes. Add the tomatoes, tomato paste and oregano. Cover and cook over low heat 1 hour. Correct seasoning.

Cook the chopped onions, celery and carrots in half the butter for 15 minutes. Cool. Add the eggs, 1 cup matzo meal, the peas, salt and pepper. Roll into small balls and dip in remaining matzo meal. Fry in the remaining butter until browned. Place on the spaghetti and pour the sauce over all. Serve with the grated Swiss cheese. Serves 6.

Baked Lasagne

> 3 tablespoons olive oil
> 1 cup minced onions
> 1 cup thinly sliced green peppers
> 2 cloves garlic, minced
> 1½ teaspoons salt
> ¼ teaspoon pepper
> ½ teaspoon oregano
> 2 tablespoons chopped parsley
> 1 can (#2½) tomatoes
> 1 can (8-ounce) tomato sauce
> ½ cup grated Dry Gouda or Edam cheese
> ½ pound lasagne (1½-inch-wide noodles)
> ¾ pound sliced Swiss cheese
> 1½ pounds cottage cheese

Heat the olive oil; cook the onions, green peppers and garlic in it for 10 minutes. Add the salt, pepper, oregano, parsley, tomatoes and tomato sauce. Cover and cook over low heat 30 minutes. Stir in 2 tablespoons Gouda cheese.

Cook the lasagne (if you can't get it, use broad noodles) as package directs. Spread ⅓ of the sauce on the bottom of a 12 x 8-inch baking dish. Arrange alternate layers of the lasagne, Swiss, cottage cheese, a sprinkling of the Gouda and spread with sauce; then repeat until all the ingredients are arranged in layers in baking dish. Cover with the remaining sauce and sprinkle top with remaining Gouda cheese.

Bake in a 350° oven 35 minutes. Serve hot. Serves 8.

Eggs Benedict

> 2 English muffins
> 1 tablespoon butter
> 4 slices smoked salmon
> 4 eggs, poached

Split the muffins and toast them. Butter and place a slice of smoked salmon on each half muffin. Carefully place a poached egg over the salmon and cover with following variation of Hollandaise sauce:

> 2 egg yolks
> ¼ teaspoon salt
> Dash cayenne pepper
> 1 tablespoon lemon juice
> ½ cup melted butter

Beat the egg yolks until thick; stir in the salt, cayenne pepper and lemon juice. Add ½ teaspoon of butter at a time, beating steadily. When half the butter is used, add the rest a little faster, still beating steadily. Serve immediately. Serves 2–4.

Corn Fritters

> 3 egg yolks
> 1¾ cups canned whole-kernel corn
> ¾ teaspoon salt
> ⅛ teaspoon pepper
> ¼ cup sifted flour
> 3 egg whites, stiffly beaten
> ½ cup salad oil
> 1 cup maple syrup

Beat the egg yolks very well; stir in the corn, salt, pepper and flour. Fold in the egg whites. Heat the oil in a skillet and drop the batter, by the tablespoon, into it. Fry until browned on both sides.

Heat the maple syrup and serve with the fritters. Serves 6.

Apple Schmarren

>   1 cup sifted flour
>   ⅛ teaspoon salt
>   2 eggs
>   1 cup milk
>   1½ cups sliced apples
>   2 tablespoons butter
>   ¼ cup sugar
>   1 tablespoon cinnamon

Sift the flour and salt into a bowl. Beat the eggs and milk together and add to flour, beating until smooth. Stir in the apples.

Melt the butter in a 9-inch skillet and pour the mixture into it.

Bake in a 350° oven 15 minutes or until set. Tear apart with 2 forks into small pieces. Sprinkle with the sugar and cinnamon. Serves 4.

NOODLES AND
KREPLACH

In this day of supermarkets, most people know little about making homemade noodles. But the very best of the packaged noodles can never equal the delicious flavor and delicate texture of a homemade noodle. Those noodles that melt in your mouth can be made at home with little difficulty.

In addition, homemade noodle dough is the basis for many other Jewish dishes, particularly *kreplach*. There is practically no limit to the variety of fillings for *kreplach*. They're particularly delicious in soup on a cold winter night.

Homemade Noodles

> 2 cups flour
> 2 eggs
> 1 tablespoon water
> ½ teaspoon salt

Place unsifted flour on a board and make a well in the center. Drop the eggs, water and salt into it. Work into the flour with one hand and knead until smooth and elastic. Roll and stretch the dough as thin as possible. The thinner it is, the better the noodles. Let the rolled dough stand until it feels dry to the touch, but don't let it get too dry. You can cut the dough into squares, strips or very narrow noodles. For narrow ones, roll up like a jelly roll and slice as thin as possible. Shake until they separate, and let dry very thoroughly. Cook the amount you want in boiling salted water or soup for about 10 minutes. Keep the balance in tightly closed jars.

Kreplach

Prepare one recipe noodle dough (see preceding recipe). Roll out but don't let it dry. Cut into 3-inch squares and place a tablespoon of one of the following mixtures on each. Fold over the dough into a triangle. Press edges together with a little water. Cook in boiling salted water or soup 20 minutes, or until they rise to the top. Drain, if cooked in water. They can then be fried or served immediately in the soup. Makes 24 or more, according to how thin you roll the dough.

FILLINGS FOR KREPLACH

MEAT:

>   1 tablespoon fat
>   ½ pound ground beef
>   ½ cup minced onions
>   ¾ teaspoon salt
>   ¼ teaspoon pepper

Heat the fat in a skillet and cook the meat and onions in it for 10 minutes, stirring frequently. Add the salt and pepper. Cool before placing in squares of dough.

KASHA:

>   1 cup minced onions
>   3 tablespoons chicken fat or butter (depending on
>   whether you serve with meat or dairy dish)
>   1½ cups cooked kasha
>   ¼ teaspoon pepper

Lightly brown the onion in the fat or butter. Stir in the kasha and pepper.

CHEESE-POTATO:

>   ½ cup minced onions
>   3 tablespoons butter
>   1½ cups mashed potatoes
>   ½ cup pot cheese
>   1 teaspoon salt
>   ⅛ teaspoon pepper
>   1 egg
>   Sour cream

Lightly brown the onions in the butter. Add the potatoes, cheese, salt, pepper and egg, beating until smooth. Serve with sour cream.

CHICKEN:

> 1½ cups ground cooked chicken
> 4 tablespoons browned minced onion
> 1 egg yolk
> ¾ teaspoon salt
> Dash pepper
> 1 tablespoon minced parsley

Blend all the ingredients together.

CHICKEN LIVER:

> ½ pound chicken livers
> ½ cup minced onions
> 2 tablespoons chicken fat
> 2 hard-cooked egg yolks
> 1 teaspoon salt
> ⅛ teaspoon pepper

Cook the livers and onions in the fat for 10 minutes, mixing frequently. Grind or chop the livers, onions, eggs, salt and pepper. Cool before placing in squares of dough.

Dairy Noodle Ring

> ½ pound cream cheese
> 2 eggs
> ¼ cup sugar
> 1 teaspoon salt
> 3 cups cooked medium noodles, drained
> 2 tablespoons dry bread crumbs
> 3 tablespoons melted butter

Cream the cheese with the eggs, sugar and salt. Stir in the noodles. Turn into a buttered 9-inch ring mold. Sprinkle with the bread crumbs and butter.

Bake in a 375° oven 40 minutes or until browned. Unmold carefully. Serves 6.

Noodles and Cheese

> 4 tablespoons butter
> ½ pound noodles, cooked and drained
> ¾ teaspoon salt
> 1½ cups pot cheese

Melt the butter in a skillet. Stir in the noodles and cook over medium heat, stirring constantly until noodles brown. Remove from the heat and add the salt and pot cheese. Mix well. Serves 3–4.

Noodles and Cabbage

> 1 tablespoon salt
> 4 cups finely shredded cabbage
> ½ cup butter or fat
> 1 teaspoon sugar
> ¼ teaspoon pepper
> 3 cups cooked broad noodles, drained

Mix the salt and cabbage together and let stand 30 minutes. Squeeze out all the liquid.

Heat the butter or fat in a deep skillet. Add the cabbage, sugar and pepper. Cook over low heat 45 minutes or until cabbage is browned. Stir very frequently. Add the noodles and toss to blend thoroughly. Serves 6–8.

Noodle-Apple Pudding

> 2 eggs
> 4 tablespoons sugar
> ¼ teaspoon salt

½ teaspoon cinnamon
1 cup grated apples
½ cup seedless raisins
4 cups cooked fine noodles, drained
3 tablespoons melted butter or fat

Beat the eggs, sugar, salt and cinnamon together. Stir in the apples, raisins, noodles and butter or fat. Turn into a greased baking dish.

Bake in a 400° oven 40 minutes or until browned. Serves 6–8.

Green Noodle Pudding

4 egg yolks
4 tablespoons chopped scallions
1½ cups cooked chopped spinach
½ pound fine noodles, cooked and drained
1 teaspoon salt
4 egg whites, stiffly beaten

Beat the egg yolks, then stir in the scallions, spinach, noodles and salt. Fold in the egg whites. Pour into a 1½-quart greased baking dish.

Bake in a 350° oven 30 minutes. Serve hot with meat or dairy dishes. Serves 4–6.

Noodle-Cheese Pudding

4 eggs
¾ cup sour cream
1 teaspoon salt
2 tablespoons sugar
2 cups cottage cheese

   5 *cups cooked fine noodles*
   4 *tablespoons dry bread crumbs*
   3 *tablespoons melted butter*

Beat the eggs, sour cream, salt and sugar together. Stir in the cheese and noodles. Turn into a buttered 2-quart baking dish or casserole. Sprinkle with the bread crumbs and butter.

Bake in a 375° oven 40 minutes. Serves 6.

Flaishig Noodle Pudding

   ½ *cup diced onions*
   ½ *cup chicken fat*
   3 *eggs*
   6 *cups cooked fine noodles, drained*

Brown the onions in the fat. Cool for 15 minutes. Beat the eggs very well and stir in the noodles, browned onions and fat. If you have any *grebenes* (page 58), add them. Turn into a greased oblong or square pan.

Bake in a 375° oven 40 minutes or until browned. Cut into squares. Serves 6–8.

Spaetzel

   2 *cups sifted flour*
   2 *eggs*
   ½ *teaspoon salt*
   2 *tablespoons cold water*

Sift the flour into a bowl and make a well in the center. Drop the eggs, salt and water into it. Work into the flour with the hand and knead until the dough leaves the sides of the bowl. Knead on a board until smooth and elastic. Divide the dough in four pieces and roll into pencil-thin strips. Cut into 1-inch pieces and drop into boiling salted water. Cook until they rise to the surface. Drain. Serve hot with meats. Serves 6–8.

Kasha Varnitchkes

> *1 cup minced onions*
> *⅓ cup chicken fat*
> *2 cups cooked kasha*
> *3 cups cooked broad noodles, drained*
> *1½ teaspoons salt*
> *¼ teaspoon pepper*

Brown the onions in the fat. Combine with the kasha, noodles, salt and pepper. If you have any *grebenes* (page 58), add some. Toss until well mixed. Serve hot. Serves 6.

Noodles with Honey and Poppy Seeds

> *½ cup honey*
> *½ cup light cream*
> *½ cup milk*
> *¼ cup poppy seeds*
> *4 cups cooked medium noodles, drained*

Cook the honey, cream, milk and poppy seeds for 5 minutes. Stir in the noodles and serve immediately. Serves 6–8.

Verenikas

> 3 cups pitted sour red cherries or blueberries
> ¾ cup sugar
> 2 teaspoons lemon juice
> 1 tablespoon cornstarch
> 1 tablespoon cold water
> Homemade noodle dough (page 95), one recipe
> 1 cup sour cream

Save any juice that oozes from the cherries (if cherries are used). Combine the fruit and juice, sugar and lemon juice in a saucepan. Mix cornstarch and water and add. Cook over low heat 10 minutes, stirring frequently. Drain, reserving the syrup.

Make the noodle dough and roll it out but do not let it dry. Cut into 3-inch circles. Place a tablespoon of the fruit on each. Fold over into half-moons and press edges together with a little water.

Cook in rapidly boiling salted water for 10 minutes or until verenikas rise to the top. Drain. Serve with the syrup and garnish with sour cream. Makes about 24.

KUGELS AND CHARLOTTES

In the Middle Ages, when Jewish cuisine was coming into existence, vegetables were available only during the harvest season. For this reason, vegetables are comparatively unimportant in most Jewish homes. In their place, *kugels* and *charlottes*, resembling puddings or pudding-soufflés, were substituted. They may be served as separate courses, as accompaniments to meats or poultry, or even as dessert if they are sweet. A potato *kugel* is so good that it may be served at any meal, and most people try to make enough so that they may be sure of leftovers, for cold *kugel* is equally delicious.

Fish Kugel

> 3 tablespoons butter
> 1 cup sliced onions
> 4 cups thinly sliced potatoes
> 3 cups cooked or canned flaked fish
> 1½ teaspoons salt
> ½ teaspoon pepper
> 2 eggs
> 1½ cups light cream

Melt the butter in a skillet. Brown the onions in it.

Arrange alternate layers of the potatoes, fish and onions in a buttered baking dish, starting and ending with the potatoes. As you arrange the layers, sprinkle the potatoes with 1 teaspoon salt and the pepper.

Beat the eggs, cream and remaining salt together and pour over the contents of baking dish.

Bake in a 350° oven 45 minutes or until firm. Serves 6.

Barley Kugel

> 1 cup pearl barley
> 4 cups boiling water
> 2 teaspoons salt
> ½ pound chopped mushrooms
> 2 onions, diced
> 2 tablespoons fat or butter
> ¼ teaspoon pepper
> 2 eggs, beaten

Stir the barley into the water. Bring to a boil; add the salt, cover and cook over low heat 45 minutes or until soft. Drain.

Brown the mushrooms and onions in the fat or butter. Add to the barley with the pepper and eggs. Taste for seasoning. Turn into a greased baking dish or casserole.

Bake in a 350° oven for 40 minutes or until browned and set. Serves 6 as a substitute for potatoes.

Rice Kugel

>4 cups boiling water
>1½ teaspoons salt
>1½ cups rice
>6 eggs
>½ cup brown sugar
>½ cup seedless raisins
>⅓ cup melted butter or fat

Boil the water, salt and rice in a covered saucepan for 10 minutes. Drain.

Beat the eggs and brown sugar together until thick. Stir in the rice, raisins and butter or fat. Turn into a greased baking dish or casserole.

Bake in a 350° oven 40 minutes or until browned. Serves 6.

Tzibbele Kugel

>6 egg yolks
>3 cups minced onions
>⅓ cup cracker or matzo meal
>1½ teaspoons salt
>¼ teaspoon pepper
>4 tablespoons melted fat or butter
>6 egg whites, stiffly beaten

Beat the egg yolks until thick. Stir in the onions, meal, salt, pepper and fat or butter. Fold in the egg whites. Turn into a greased 2-quart casserole.

Bake in a 350° oven for 40 minutes or until set. Serves 6–8.

Kraut Kugel

> 5 cups finely shredded cabbage
> 2 teaspoons salt
> ⅓ cup butter or fat
> ½ cup boiling water
> 1½ cups cubed white bread
> ⅓ cup potato flour
> ¼ cup seedless white raisins
> ¾ cup sliced blanched almonds
> 2 tablespoons sugar
> 4 eggs

Cook the cabbage and salt in the butter over low heat for 30 minutes, stirring frequently. Cool.

Pour the water over the bread and squeeze dry. Mash. Add the potato flour, raisins, almonds and sugar. Separate the eggs and add the yolks and cabbage. Mix until smooth.

Beat the egg whites until stiff but not dry and fold into the mixture.

Turn into a greased 2-quart casserole. Bake in a 350° oven for 40 minutes, or until set. Serves 6.

Potato Kugel

> 3 eggs
> 3 cups grated, drained potatoes

⅓ cup potato flour
½ teaspoon baking powder
1½ teaspoons salt
⅛ teaspoon pepper
3 tablespoons grated onion
4 tablespoons melted butter or fat

Beat the eggs until thick. Stir in the potatoes, potato flour, baking powder, salt, pepper, onion and butter or fat.

Turn into a greased 1½-quart baking dish or casserole. Bake in a 350° oven until browned, about 1 hour. Serve hot. Serves 6–8.

Noodle Kugel

3 eggs
4 tablespoons brown sugar
⅛ teaspoon nutmeg
4 cups cooked broad noodles
½ cup seedless white raisins
½ cup sliced blanched almonds
1 tablespoon lemon juice
4 tablespoons melted butter or chicken fat
2 tablespoons bread crumbs

Beat the eggs and brown sugar until fluffy. Add the nutmeg, noodles, raisins, almonds, lemon juice and melted butter or fat. Turn into a well-greased ring mold or baking dish. Sprinkle with the bread crumbs. Bake in a 375° oven 50 minutes or until browned.

Serve with meat or poultry dishes or as a dessert with a sweet fruit sauce. Serves 6–8.

Onion Charlotte

>   4 cups diced onions
>   2 tablespoons butter
>   ½ cup milk
>   3 egg yolks
>   1 teaspoon salt
>   ⅛ teaspoon pepper
>   1 cup light cream
>   ⅓ cup bread crumbs

Cook the onions, butter and milk over low heat for 10 minutes. Drain and turn into a buttered 9-inch pie plate.

Beat the egg yolks, salt, pepper and cream together. Pour over the onions and sprinkle with the bread crumbs.

Bake in a 350° oven for 35 minutes or until set. Serve hot. Serves 6.

Potato-Carrot Charlotte

>   1 cup grated carrots
>   ¾ cup water
>   3 cups grated, drained potatoes
>   3 egg yolks, beaten
>   4 tablespoons cracker meal
>   1½ teaspoons salt
>   1 teaspoon sugar
>   ½ teaspoon ginger
>   4 tablespoons melted butter or fat
>   3 stiffly beaten egg whites

Cook the carrots in the water for 15 minutes. Let cool in the water.

Mix the potatoes, egg yolks, cracker meal, salt, sugar,

ginger, butter or fat and the undrained carrots. Fold in the egg whites. Turn into a greased 2-quart baking dish. Bake in a 350° oven 1 hour. Serve hot. Serves 6–8.

Apple Charlotte (Passover)

    3 egg yolks
    ⅔ cup sugar
    Dash salt
    2 cups grated apples
    ⅓ cup matzo meal
    2 teaspoons grated lemon rind
    1 tablespoon slivovitz (plum brandy)
    3 egg whites, stiffly beaten
    4 tablespoons ground pecans

Beat the egg yolks, sugar and salt until thick and lemon colored. Stir in the apples, matzo meal, lemon rind and slivovitz. Fold in the egg whites. Turn into a greased 8-inch spring form. Sprinkle the nuts on top.

Bake in a 350° oven 35 minutes or until brown and firm. Cool before removing sides of pan. Serves 6–8.

Matzo-Farfel Charlotte (Passover)

    2 cups matzo farfel
    3 egg yolks
    ⅔ cup sugar
    ¼ cup sweet red wine
    ½ teaspoon salt
    2 teaspoons grated orange rind
    2 tablespoons salad oil
    3 egg whites, stiffly beaten

Soak the farfel in cold water for a few minutes, then drain and crush to a paste. Beat the egg yolks and sugar together until thick. Stir in the wine, salt, orange rind, oil and farfel. Fold in the egg whites. Turn into a greased 1½-quart baking dish.

Bake in a 350° oven 30 minutes or until browned. Serves 6–8.

VEGETABLES

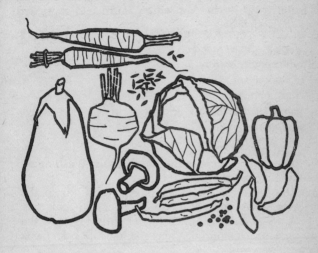

Sweet-and-Sour Green Beans

    1 cup boiling water
    2 packages frozen green beans
    1 teaspoon salt
    ⅛ teaspoon pepper
    1 bay leaf
    3 cloves
    2 tablespoons sugar
    4 tablespoons vinegar
    3 tablespoons butter or fat

Combine the water, beans, salt, pepper, bay leaf and cloves in a saucepan. Cover and cook over low heat 3 minutes less than package directs. Drain and add the sugar, vinegar and butter or fat. Cook 5 minutes, stirring frequently. Remove bay leaf. Serves 6.

Braised Kale

    3 pounds kale
    4 tablespoons butter or fat
    1 clove garlic, minced
    3 tablespoons water
    1 teaspoon salt
    ⅛ teaspoon pepper
    1 teaspoon potato flour
    2 teaspoons lemon juice

Wash the kale and remove the tough veins. Wash again in several changes of water.

Melt the butter or fat in a saucepan and stir in the garlic, kale, water, salt and pepper. Cover and cook over low heat 15 minutes or until tender. Mix together the potato flour and lemon juice and stir into the kale. Cook 2 minutes. Serves 4–5.

Bean Panache

    1 onion, chopped
    2 tablespoons butter
    ½ pound green beans, cooked and drained
    ½ pound wax beans, cooked and drained
    ½ pound lima beans, cooked and drained
    1 tablespoon chopped parsley

Cook the onion in the butter for 10 minutes. Toss with the green, wax and lima beans. Add salt and pepper to taste and sprinkle with the parsley. Serves 4–6.

Prunes and Beets

> 1 pound unsweetened prunes
> 3 cups undrained julienne beets
> ½ teaspoon salt
> 4 tablespoons sugar
> 2 teaspoons vinegar

Combine all the ingredients in a saucepan. Cover and cook over low heat 25 minutes or until prunes are tender. Delicious when served as an accompaniment to roast meat or poultry. Serves 4–6.

Creamed Beets

> 3 cups grated beets
> 1 tablespoon lemon juice
> ½ cup water
> 1½ tablespoons potato flour
> 1 cup light cream
> 1 teaspoon salt
> 2 tablespoons butter

Cook the beets, lemon juice and water for 15 minutes or until the beets are tender. Drain.

Mix the potato flour and cream in a saucepan. Cook over low heat for 5 minutes, stirring constantly. Stir in the beets, salt and butter. Cook 5 minutes. Serves 6.

Rice-Stuffed Cabbage

> 1 head cabbage
> 2 eggs
> 4 tablespoons grated onion

2 cups half-cooked rice
2 teaspoons salt
1½ cups seedless raisins
4 tablespoons butter
2 onions, sliced
1 tablespoon potato flour
2 cups boiling water
2 tablespoons lemon juice
2 teaspoons sugar
½ teaspoon cinnamon
½ cup sour cream

Pour boiling water over the cabbage and let it soak 10 minutes. Remove the top 18 leaves (if not large enough use 24).

Beat the eggs and gradually add the grated onion, rice, 1 teaspoon salt and half the raisins. Place a heaping tablespoon of the mixture on each leaf. Carefully roll up.

Melt the butter in a casserole. Lightly brown the onions and cabbage rolls in it. Sprinkle with the potato flour, then add the boiling water, lemon juice, sugar, cinnamon and the remaining salt and raisins. Bake in a 350° oven 1¼ hours, basting frequently. Stir in the sour cream just before serving. Serves 6–8.

Dairy-Style Cabbage

6 cups shredded cabbage
4 tablespoons butter
1½ teaspoons salt
1 tablespoon lemon juice
2 tablespoons sugar
1 egg
1 cup sour cream

Cook the cabbage in the butter over low heat for 45 minutes, stirring frequently. Stir in the salt, lemon juice and sugar. Cook 5 minutes longer.

Beat the egg and sour cream together and add to the cabbage, mixing steadily until it begins to thicken. Serves 4–6.

Sweet-and-Sour Red Cabbage

6 cups shredded red cabbage
½ cup grated apple
¾ cup water
1 teaspoon salt
1 tablespoon flour
4 tablespoons vinegar
1½ tablespoons sugar
2 tablespoons butter or fat

Cook the cabbage, apple, water and salt over low heat 15 minutes. Mix the flour and vinegar until smooth and add to the cabbage with the sugar and butter or fat. Cook 15 minutes longer. Serves 6.

Honeyed Carrots

3 tablespoons butter or fat
4 cups sliced carrots
3 tablespoons orange juice
1½ teaspoons salt
¼ teaspoon ginger
4 tablespoons honey

Combine all the ingredients in a saucepan. Cover and cook over low heat 25 minutes, stirring occasionally. Serves 6.

Fried Eggplant

>   1 large eggplant
>   2 tablespoons lemon juice
>   2½ teaspoons salt
>   1 egg
>   ⅛ teaspoon pepper
>   1 cup dried bread crumbs
>   ⅓ cup fat or butter

Peel the eggplant and slice thin. Sprinkle with the lemon juice and 1½ teaspoons salt. Let stand 30 minutes. Rinse with cold water. Drain.

Beat the egg, pepper and remaining salt together. Dip the eggplant slices in the egg mixture and then in the bread crumbs. Brown on both sides in hot fat or butter. Serves 4–6.

Kasha

>   1½ cups buckwheat groats
>   1 egg, beaten
>   2½ cups boiling water
>   1½ teaspoons salt

You can buy fine, medium or coarse groats; use whichever type you prefer. Mix the groats and egg in a saucepan over low heat until the grains separate. Add the water and salt. Cover and cook over low heat 15 minutes. All the water should be absorbed; if not, drain thoroughly.

Serve with melted fat or butter or use in other recipes. Serves 6.

Baked Sweet Lima Beans

1½ *cups dried lima beans*
2 *teaspoons salt*
3 *tablespoons butter or fat*
½ *cup honey*

Soak the beans overnight in water to cover. Drain. Add fresh water to cover.

Cover and bring to a boil; cook over low heat 1½ hours, adding the salt after 1 hour of cooking. Drain.

Combine the beans, butter or fat and honey in a baking dish. Bake in a 350° oven 1 hour, stirring frequently. Serves 6–8.

Baked Limas in Chili Sauce

1½ *cups dried lima beans*
2 *teaspoons salt*
2 *teaspoons sugar*
1 *cup chili sauce*
2 *tablespoons melted butter or fat*
1 *bay leaf*

Soak the beans overnight in water to cover. Drain. Add fresh water to cover, and bring to a boil. Add the salt and sugar. Cook over low heat 1½ hours. Drain.

Combine the beans, chili sauce, butter or fat and bay leaf in a baking dish. Bake in a 350° oven 1 hour. Remove bay leaf. Serves 6.

Honeyed Nahit

> 2 cups chickpeas
> 2 teaspoons salt
> 2 tablespoons melted fat or butter
> ½ cup water
> ½ cup honey

Wash the chickpeas and soak overnight in water to cover. Drain. Cover with fresh water; bring to a boil and cook over medium heat 1½ hours. Drain.

Mix the chickpeas, salt, fat or butter, ½ cup water and the honey in a baking dish.

Bake in a 350° oven 30 minutes. Serves 6.

Nahit and Rice

> ¾ cup chickpeas
> ½ cup raw rice
> 1 teaspoon salt
> 2 cups boiling water
> ½ cup honey

Wash the chickpeas and soak overnight in water to cover. Cover with fresh water; bring to a boil and cook over low heat 2 hours. Drain.

Stir the rice and salt into the 2 cups boiling water. Cover and cook over low heat 15 minutes. Turn the undrained rice into a casserole and stir in the honey and chickpeas.

Bake in a 350° oven for 35 minutes. Serves 4–6.

Dairy Stuffed Peppers

6 green peppers
5 tablespoons butter
2 cups diced onions
1 pound mushrooms, chopped
2 cups half-cooked rice
2½ teaspoons salt
⅛ teaspoon pepper
1 egg
2 tablespoons potato flour
1 cup water
2 cups canned tomatoes
½ cup sour cream

Cut a 1-inch piece from the stem ends of the peppers. Scoop out the seeds and fibers carefully. Cook the peppers in boiling water for 5 minutes, then drain.

Melt half the butter in a skillet and brown half the onions and all the mushrooms in it. Stir in the rice, 1½ teaspoons salt, the pepper and the egg. Stuff the peppers with the mixture.

Mix the potato flour and water in a saucepan. Stir in the tomatoes, remaining salt and onions. Arrange the peppers in it, open end up. Cover and cook over low heat 1 hour, basting frequently. Stir in the sour cream a few minutes before serving. Serves 6.

Hot Sauerkraut

2 tablespoons fat or butter
¾ cup diced onions
1½ pounds sauerkraut, drained
¼ teaspoon pepper

> 2 teaspoons paprika
> 1 teaspoon sugar
> ½ cup water

Melt the fat in a saucepan and brown the onions lightly. Add the sauerkraut, pepper, paprika, sugar and water. Cover loosely and cook over low heat 1 hour. Serves 4–6.

Baked Acorn Squash

> 2 acorn squash
> ¼ cup melted butter or fat
> ¼ cup corn syrup
> ½ teaspoon salt

Cut the squash in half lengthwise and scoop out the seeds and fibers. Prick the flesh in several places with a fork. Place in a baking pan cut side down. Bake in a 375° oven 30 minutes. Mix together the butter or fat, corn syrup and salt. Turn the squash over and divide the mixture among the four halves. Bake 30 minutes longer, basting frequently. Serves 4.

Squash and Rice

> 3 pounds yellow squash
> 1 teaspoon salt
> 1 tablespoon rice
> 2 teaspoons sugar
> 3 tablespoons butter
> ½ cup light cream

Peel the squash and slice it very thin. Cook the squash in a saucepan over low heat until the liquid from the squash covers the bottom of pan. Add the salt and rice.

Cover loosely and cook over low heat 50 minutes. Watch carefully while cooking and add a little water if necessary. Mash the squash and stir in the sugar, butter and cream. Taste for seasoning. Serves 6.

Candied Sweet Potatoes

  6 sweet potatoes
  ¼ cup butter or fat
  ½ cup dark corn syrup
  2 tablespoons orange juice
  ¼ cup brown sugar

Cook the unpeeled potatoes in boiling water for 15 minutes. Peel and cut in half lengthwise. Heat the butter or fat in a baking dish and add the corn syrup, orange juice, sugar and sweet potatoes.

Bake in a 350° oven 45 minutes or until tender, basting frequently. Serves 6–8.

Sweet Potatoes Royal

  1 cup dried apricots
  1 cup brown sugar
  2 pounds sweet potatoes, cooked and peeled
  ¼ cup melted butter or fat
  ½ cup sliced, blanched almonds

Wash the apricots and soak in 2 cups of water for 2 hours. Bring to a boil and cook over low heat 20 minutes or until tender. Stir in the sugar.

Slice the potatoes ½ inch thick. Arrange layers of the potatoes and undrained apricots in a casserole. Pour the butter or fat over the top.

Bake in 375° oven 35 minutes, basting twice. Sprinkle with the almonds and bake 10 minutes longer. Serves 8.

Sweet Potatoes Hawaiian

¼ *cup butter or fat*
6 *sweet potatoes, peeled and cut in half lengthwise*
1 *cup brown sugar*
1½ *cups undrained crushed pineapple*

Melt the butter or fat in a skillet; brown the potatoes in it. Add the brown sugar and pineapple. Cook over low heat 45 minutes or until tender. Turn the potatoes frequently. Serves 6–8.

TZIMMES

A *tzimmes* means a fuss or excitement. But making a *tzimmes* does not mean a great deal of work; most *tzimmes* dishes are easily prepared. Actually, it's almost any combination of meat or vegetables or fruits, limited only by the imagination of the cook. *Tzimmes* is usually served with the main course instead of green vegetables or potatoes. In many families, they serve vegetables, potatoes and *tzimmes,* because what does one have to do with the other?

An important thing to remember about *tzimmes* cookery is that it should be cooked as long as possible in order to blend the various flavors of the ingredients. A quick *tzimmes* is possible and it won't be bad, but for the real thing, cook your *tzimmes* slowly and gently over low heat.

Dairy Tzimmes

1 pound prunes
4 cups boiling water
1 cup farfel
1 teaspoon salt
2 tablespoons lemon juice
⅓ cup honey
4 tablespoons butter

Wash the prunes and soak in the water for 1 hour. Bring to a boil and add the farfel, salt, lemon juice, honey and butter.

Place mixture in baking dish. Cover and bake in a 375° oven 45 minutes, removing the cover for the last 15 minutes. Serves 6.

Carrot and Apple Tzimmes

4 cups grated carrots
1 tablespoon fine barley
¾ cup grated apples
3 tablespoons butter or fat
½ cup water
½ teaspoon salt
2 teaspoons sugar
¼ teaspoon nutmeg

Combine all the ingredients in a saucepan. Cover and cook over low heat 2 hours, or until the barley is soft. Watch carefully while cooking and add more water if necessary. Serves 6.

Mixed-Fruit Tzimmes

½ pound dried apricots
½ pound dried pears
½ pound unsweetened prunes
3 pounds brisket
2 teaspoons salt
¼ teaspoon pepper
3 carrots, quartered
6 thin slices lime or lemon
2 cups orange juice
4 cups water
4 tablespoons honey

Wash the apricots, pears and prunes and soak in cold water to cover for 1 hour. Drain.

Brown the meat over medium heat in a Dutch oven or casserole. Sprinkle with the salt and pepper. Arrange the fruit, carrots and lime around it. Mix the orange juice, water and honey together and pour over all. Cover and bake in a 350° oven 3 hours. Remove the cover and increase heat to 400°. Bake 1 hour longer, adding a little water if necessary. Serves 6–8.

Sweet-Potato-and-Prune Tzimmes

1½ pounds prunes
3 cups boiling water
2 tablespoons fat
3 pounds brisket
2 onions, diced
1½ teaspoons salt
¼ teaspoon pepper

3 *sweet potatoes, peeled and quartered*
½ *cup honey*
2 *cloves*
½ *teaspoon cinnamon*

Wash prunes and let soak in the boiling water ½ hour.

Melt the fat in a Dutch oven. Cut the beef in 6 or 8 pieces and brown with the onions. Sprinkle with the salt and pepper. Cover and cook over low heat 1 hour. Add the undrained prunes, sweet potatoes, honey, cloves and cinnamon. Replace cover loosely and cook over low heat 2 hours. Serves 6–8.

Grated-Potato Tzimmes

1 *pound unsweetened prunes*
3 *cups water*
8 *potatoes, peeled*
⅓ *cup brown sugar*
2 *tablespoons lemon juice*
1½ *teaspoons salt*
3 *tablespoons potato flour*
2 *tablespoons melted fat*
⅛ *teaspoon pepper*

Wash the prunes and soak in the water 1 hour. Bring to a boil.

Slice 5 potatoes into a 2-quart casserole and pour the undrained prunes over them. Add the brown sugar, lemon juice and 1 teaspoon salt. Cover and bake in a 350° oven 1 hour.

Grate the remaining potatoes and add the potato flour, fat, pepper and remaining salt. Remove the cover of the casserole and spread the mixture over the top. Replace cover and bake additional 1¼ hours, removing the cover for the last 45 minutes. Serves 6–8.

Sweet-Potato Tzimmes

> 2 tablespoons fat
> 3 pounds brisket
> 4 cups boiling water
> 1½ teaspoons salt
> ¼ teaspoon pepper
> ¼ teaspoon nutmeg
> 8 carrots, sliced
> 4 sweet potatoes, peeled and cut in quarters
> ⅓ cup brown sugar

Melt the fat in a Dutch oven or casserole and brown the beef. Add the water, salt, pepper and nutmeg. Cover and cook over low heat 1½ hours. Add the carrots, sweet potatoes and brown sugar. Cover and bake in a 350° oven 2 hours, removing the cover for the last half hour. Serves 6–8.

KNISHES, PIROSHKI AND BLINTZES

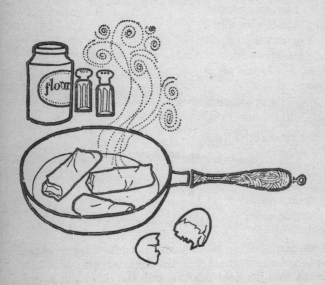

The Jewish style of cooking places considerable emphasis upon noodles, pancakes and dough. Like the Italian or Chinese cuisine, it is difficult to imagine a Jewish meal that doesn't include some kind of dough preparation.

Knishes, always a great favorite, have become enormously popular in recent years as cocktail-party snacks. Of course, the modern version is daintier than the old, familiar *knish.* But it still tastes the same despite its streamlined appearance.

Piroshki are most often served with soups and make a welcome change from the bread or crackers which are inevitably served with soup. Incidentally, there is no reason why *piroshki* can't be served as cocktail-party snacks just as *knishes* are.

Blintzes are very similar to *crêpes*, the French pancakes. *Blintzes* are one of the mainstays of Jewish cookery, and are frequently a main course of dairy meals. Cheese *blintzes* are still the leading favorite, but have you ever had cherry or huckleberry blintzes?

Knishes

DOUGH:

> 2½ cups sifted flour
> 1 teaspoon baking powder
> ½ teaspoon salt
> 2 eggs
> ⅔ cup salad oil
> 2 tablespoons water

Sift the flour, baking powder and salt into a bowl. Make a well in the center and drop the eggs, oil and water into it. Work into the flour mixture with the hand and knead until smooth.

There are two ways to fill the knishes. In either case, divide the dough in two and roll as thin as possible. Brush with oil. Now you can spread the filling on one side of the dough and roll it up like a jelly roll. Cut into 1½-inch slices. Place in an oiled baking sheet cut side down. Press down lightly to flatten. Or you can cut the rolled dough in 3-inch circles. Place a tablespoon of the filling on each; draw the edges together and pinch firmly. Place on an oiled baking sheet, pinched edges up. In either case, bake in a 375° oven 35 minutes or until browned. Makes about 24.

FILLINGS FOR KNISHES

POTATO:

>   1 cup chopped onions
>   6 tablespoons chicken fat or butter
>   2 cups mashed potatoes
>   1 egg
>   1 teaspoon salt
>   ¼ teaspoon pepper

Brown the onions in the fat or butter. Beat in the potatoes, egg, salt and pepper until fluffy.

CHEESE:

>   1½ cups diced scallions or onions
>   4 tablespoons butter
>   2 cups pot cheese
>   1 egg
>   1½ teaspoons salt
>   ⅛ teaspoon pepper
>   2 tablespoons sour cream

Scallions are better than onions for this, so try to get them. Brown the scallions in the butter and beat in the cheese, egg, salt, pepper and sour cream until smooth.

MEAT:

>   ½ cup minced onions
>   2 tablespoons chicken fat
>   1½ cups ground cooked meat
>   ½ cup cooked rice
>   1 egg

1 teaspoon salt
¼ teaspoon pepper

Lightly brown the onions in the fat. Add the meat, rice, egg, salt and pepper, mixing until smooth.

CHICKEN:

1½ cups ground cooked chicken
¾ cup mashed potatoes
1 egg
1 teaspoon salt
¼ teaspoon pepper

Mix all the ingredients until smooth.

Potato-Dough Knishes

¾ cup minced onions
6 tablespoons chicken fat or butter
4 cups mashed potatoes
½ cup potato flour
3 eggs
1 teaspoon salt
¼ teaspoon pepper

Brown the onions in 4 tablespoons chicken fat or butter. Cool.

Knead together the remaining fat or butter, the potatoes, potato flour, eggs, salt and pepper. Break off pieces (about 2 inches long) and flatten slightly. Place a teaspoon of browned onions on each and cover by pinching the edges together. Place on a greased baking sheet.

Bake in a 375° oven 25 minutes. Makes about 20.

Blintzes

BLINTZE BATTER:
>   *3 eggs*
>   *1 cup milk or water*
>   *½ teaspoon salt*
>   *2 tablespoons salad oil*
>   *¾ cup sifted flour*
>   *Butter or oil for frying*

Beat the eggs, milk, salt and salad oil together. Stir in the flour.

Heat a little butter or oil in a 6-inch skillet. Pour about 2 tablespoons of the batter into it, tilting the pan to coat the bottom. Use just enough batter to make a very thin pancake. Let the bottom brown, then carefully turn out onto a napkin, browned side up. Make the rest of the pancakes.

Spread 1 heaping tablespoon of any of the fillings along one side of the pancake. Turn opposite sides in and roll up like a jelly roll.

You can fry the blintzes in butter or oil or bake them in a 425° oven until browned. Makes about 18.

Serve dairy blintzes with sour cream.

Sour-Cream Batter for Blintzes

>   *1 egg*
>   *¼ cup milk*
>   *¾ cup sour cream*
>   *⅛ teaspoon salt*
>   *1 cup sifted flour*
>   *Butter for frying*

Beat together the egg, milk, sour cream and salt. Stir in the flour, mixing until smooth.

Heat some butter in a 7-inch skillet. Pour about 2 tablespoons of the batter into it, tilting the pan to spread the batter evenly. Fry until browned and turn to brown other side.

Place a heaping tablespoon of one of the fillings on each pancake. Tuck in the opposite sides and roll up. Arrange in a buttered baking dish and bake in a 450° oven 10 minutes.

This batter makes a rich pancake, and is more suitable for sweet fillings. Makes about 16.

FILLINGS FOR BLINTZES

APPLE:

>   *1 egg white*
>   *1½ cups finely chopped apples*
>   *4 tablespoons sugar*
>   *½ teaspoon cinnamon*
>   *3 tablespoons brown sugar*
>   *3 tablespoons melted butter*

Beat the egg white until it begins to stiffen. Fold in the apples, sugar and cinnamon. Fill the pancakes and arrange in a buttered baking pan. Sprinkle with the brown sugar and butter.

Bake in a 400° oven 20 minutes. Makes about 18.

VEGETABLE:

>   *½ cup shredded cabbage*
>   *½ cup grated carrots*
>   *½ cup finely sliced green pepper*
>   *¾ cup diced onions*

> 3 tablespoons butter or oil
> 1 teaspoon salt
> Dash cayenne pepper

Cook the cabbage, carrots, green pepper and onions in the butter or oil for 10 minutes, stirring occasionally. Season with the salt and cayenne.

MEAT:

> 2 cups ground cooked meat
> 2 tablespoons grated onion
> 1 egg
> 1 teaspoon salt
> ¼ teaspoon pepper
> 2 tablespoons minced parsley

Mix all the ingredients together.

CHEESE:

> 2 cups drained cottage cheese
> 1 egg yolk
> ¾ teaspoon salt
> 1 tablespoon melted butter
> 2 tablespoons sugar (optional)
> 1 teaspoon lemon juice (optional)

Beat the cheese, egg yolk, salt and butter together. Add the sugar or lemon juice if you like—some people like them sweet, some don't.

BLUEBERRY:

> 1½ cups blueberries
> 3 tablespoons sugar
> 1 tablespoon cornstarch
> ⅛ teaspoon nutmeg

Toss all the ingredients together.

Piroshki

DOUGH:

>2 cups sifted flour
>½ teaspoon salt
>¾ cup shortening
>1 egg yolk
>4 tablespoons ice water

Sift the flour and salt together. Work in the shortening with the hand. Beat the yolk and water together and add to the previous mixture. Toss lightly and form into a ball.

Roll out the dough ⅛ inch thick and cut into 3-inch circles. Use a tablespoon of one of the following fillings for each. Chill. Fold over into a half-moon and press the edges together with a little water. Arrange on a greased baking sheet.

Bake in a 400° oven 15 minutes or until browned. Makes about 24.

FILLINGS FOR PIROSHKI

POTATO:

>½ cup minced onions
>4 tablespoons chicken fat or butter
>2 cups mashed potatoes
>1 egg yolk
>1 teaspoon salt
>Dash cayenne pepper

Lightly brown the onions in the fat or butter. Stir in the potatoes, egg yolk, salt and cayenne pepper. Cool 10 minutes.

LIVER:

> ½ cup minced onions
> ¼ pound mushrooms, chopped
> 6 tablespoons chicken fat
> ½ pound chicken livers
> 2 tablespoons minced parsley
> 1½ teaspoons salt
> ¼ teaspoon pepper

Brown the onions and mushrooms in half the fat. Remove onions and mushrooms and reserve. Melt remaining fat in the same pan and cook the livers in it 10 minutes. Chop together, until very smooth, the onions, mushrooms, livers, parsley, salt and pepper. Cool.

FISH:

> 1½ cups cooked flaked fish
> 4 tablespoons grated onion
> 2 hard-cooked egg yolks
> 4 tablespoons sour cream
> 1 teaspoon salt
> ⅛ teaspoon pepper

Mix all the ingredients together.

MUSHROOM:

> 1 pound mushrooms, sliced
> ¾ cup minced onions
> 3 tablespoons butter
> 2 hard-cooked egg yolks
> 3 tablespoons sour cream
> 4 tablespoons bread crumbs
> 1½ teaspoons salt
> Dash cayenne pepper

Cook the mushrooms and the onions in the butter for 15 minutes. Chop the mushroom mixture and eggs together. Stir in the sour cream, bread crumbs, salt and cayenne pepper.

CHEESE:

> ½ *pound pot cheese*
> ¼ *pound cream cheese*
> 2 *tablespoons sour cream*
> 1 *egg yolk*
> 1 *teaspoon salt*
> 1 *tablespoon sugar (optional)*

Beat all the ingredients until smooth. Use the sugar or not, depending on whether you like a sweet filling. Serve with sour cream.

PRUNE:

> ¼ *cup honey*
> ¾ *cup orange juice*
> 2 *teaspoons lemon juice*
> 1 *pound unsweetened pitted prunes*
> 1 *tablespoon grated orange rind*

Cook the honey, orange juice and lemon juice for 5 minutes. Add the prunes and cook over low heat 15 minutes, stirring occasionally. Drain and chop the prunes. Add the orange rind. Cool before filling the squares.

Beet Salad

2 cups grated cooked beets
4 tablespoons grated horseradish
2 teaspoons sugar
1 teaspoon salt
2 teaspoons vinegar
2 tablespoons salad or olive oil

Mix all the ingredients together and chill for 2 hours.
Serves 6–8.

Cole Slaw

> 4 cups shredded cabbage
> ½ cup grated carrots
> 1 cup thinly sliced green pepper
> ¼ cup water
> ¾ cup cider vinegar
> ½ cup mayonnaise
> 1 tablespoon sugar
> 2 teaspoons salt
> ¼ teaspoon pepper
> ½ teaspoon celery seed
> ¼ teaspoon mustard seed

Toss together the cabbage, carrots and green pepper. Gradually stir the water and vinegar into the mayonnaise. Add the sugar, salt, pepper, celery and mustard seeds. Pour the dressing over the vegetables and toss until well mixed. Chill 30 minutes before serving. Serves 6–8.

Wilted Red-Cabbage Slaw

> 2 pounds red cabbage
> 3 cups boiling water
> 1 tablespoon salt
> 4 tablespoons salad or olive oil
> ⅓ cup lemon juice
> ¼ teaspoon pepper
> 2 teaspoons sugar
> 4 tablespoons minced onion
> ½ cup grated apple

Shred the cabbage and combine with the boiling water and salt. Let stand 10 minutes. Drain.

Mix the oil, lemon juice, pepper and sugar together. Add to the cabbage with the onion and apple. Toss until well mixed. Chill for ½ hour before serving. Serves 6–8.

Cucumbers in Dill

> 4 cucumbers
> 1 cup boiling water
> ¾ cup sour cream
> ¼ cup lemon juice
> 3 tablespoons minced dill
> 1½ teaspoons salt
> ⅛ teaspoon pepper
> 1 teaspoon sugar

Peel the cucumbers and slice very thin. Pour the boiling water over them and let stand for 5 minutes. Drain and plunge into ice water. Drain again.

Mix together the sour cream, lemon juice, dill, salt, pepper and sugar. Pour over the cucumbers, tossing until well mixed. Chill for 30 minutes before serving. Serves 6–8.

Cucumber Salad

> 3 cucumbers
> 2 teaspoons salt
> ½ cup cider vinegar
> 2 tablespoons cold water
> 1 teaspoon sugar
> ⅛ teaspoon pepper
> 2 scallions, sliced thin

Peel the cucumbers and slice very thin. Sprinkle with the salt and set aside for 20 minutes. Drain well.

Mix together the vinegar, water, sugar and pepper. Add to the cucumber with the scallions. Mix well and chill for 2 hours before serving. Serves 4–6.

Pepper Salad

> 8 green or red peppers
> 1½ teaspoons salt
> ¼ teaspoon pepper
> ½ teaspoon garlic powder
> 1 bay leaf
> 1 cup cider vinegar
> ¼ cup water

Cut the peppers in half, discarding the seeds and fibers. Pierce each pepper with a fork and hold it over a flame or heat until the skin browns and blisters. Rub the skin off, then slice the peppers.

Combine the salt, pepper, garlic powder, bay leaf, vinegar and water in a saucepan. Bring to a boil and pour over the peppers. Chill for a few hours. Remove the bay leaf. This salad will keep 2 days. Serves 6–8.

Pickled Vegetable Salad

> ¾ cup finely diced green peppers
> ¾ cup minced onions
> 3 cups coarsely grated beets
> 3 cups grated cabbage

2 cups cider vinegar
1 tablespoon salt
½ cup sugar
2 tablespoons mustard seed
1½ tablespoons celery seed

Combine all the ingredients in a saucepan. Bring to a boil and cook over low heat 10 minutes, mixing occasionally.

Turn into two 2-quart sterile jars and seal at once. Chill 36 hours before serving. Serves 12–16.

Mock Seafood Salad

½ cup mayonnaise
½ cup sour cream
3 tablespoons cider vinegar
2 tablespoons anchovy paste
½ teaspoon salt
1 teaspoon Worcestershire sauce
¼ teaspoon pepper
4 tablespoons minced scallions
2 tablespoons minced parsley
1 clove garlic, minced
1 head lettuce, shredded
4 endive, sliced
1 can tuna fish, drained and flaked

Mix together the mayonnaise, sour cream, vinegar, anchovy paste, salt, Worcestershire sauce, pepper, scallions, parsley and garlic. Add the lettuce, endive and tuna fish. Toss and serve cold with sliced tomatoes. Serves 4.

Potato Salad

> 2 pounds potatoes
> ¼ cup minced onions
> 1½ teaspoons salt
> ¼ teaspoon pepper
> ½ cup mayonnaise
> 2 tablespoons vinegar

Buy small potatoes and cook them in boiling water about 20 minutes or until tender but firm. Peel and cube the potatoes. Add the onions, salt and pepper. Mix together the mayonnaise and vinegar; pour over the salad. Toss until well blended. Serve hot or cold. Serves 4–6.

Spring Salad

> 1 cucumber, peeled and diced
> 8 radishes, sliced
> 6 scallions, sliced
> 1 tomato, diced
> 1 pound cottage or pot cheese
> 1 teaspoon salt
> 1 pint sour cream

Lightly mix all the ingredients together. Serve with buttered rye bread.

If you like, arrange six portions of the cheese on individual plates and flatten down the centers. Mix the cucumbers, radishes, scallions and tomato together and place on the cheese. Garnish each portion with sour cream. Serves 6.

Lettuce with Sour-Cream Dressing

> ½ *cup sour cream*
> ¼ *cup cider vinegar*
> 1 *teaspoon salt*
> ¼ *teaspoon pepper*
> 6 *scallions, sliced*
> 1 *head lettuce*

Mix together the sour cream, vinegar, salt, pepper and scallions.

Tear (don't cut) the lettuce into small pieces and pour the dressing over it. Toss. Serves 4.

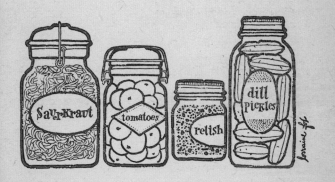

Homemade Sauerkraut

> *12 pounds cabbage*
> *4 tablespoons salt*
> *3 green apples*

A wooden keg is best for sauerkraut, but if you don't have one, use a crock or sterile glass jar.

Shred the cabbage finely (reserving some large leaves) and mix with the salt. Peel and shred the apples. Arrange layers of the cabbage and apples, pressing down very firmly after each layer. Cover with the reserved cabbage leaves and then with a piece of cheesecloth. Place a plate or board over the cheesecloth and weight it down. Keep in a cool place for 2 weeks, then taste to see if it is pickled enough. If it isn't, wash the cheesecloth and board and replace for another 5 days. Taste again.

Cucumber Relish

5 cucumbers, peeled and chopped
1½ pounds tomatoes, diced
2 cups finely shredded cabbage
2 green peppers, minced
2 onions, minced
1¼ cups vinegar
1½ cups sugar
2 teaspoons salt
1 tablespoon mustard seed
½ tablespoon celery seed

Mix all the ingredients together very thoroughly. Pack into three 3-quart sterile jars. Chill for 36 hours before using.

Pickled Tomatoes

30 green tomatoes
½ cup salt
2 quarts water
1 cup white vinegar
4 cloves garlic
3 bay leaves
¼ teaspoon whole peppercorns
1 teaspoon pickling spice
10 sprigs dill

Buy even-sized, very firm tomatoes. Wash and dry them. Arrange in a crock or glass jar.

Bring the salt and water to a boil. Cool and add the vinegar, garlic, bay leaves, peppercorns and pickling spice. Pour over the tomatoes. Arrange the dill on top.

Be sure the liquid covers the tomatoes; if not add more salted water. Cover with a plate or wooden board to weight it down if crock is used. Keep in a cool place for a week.

Pumpkin Pickle

> *4 cups peeled, cubed pumpkin*
> *¾ cup white vinegar*
> *1 cup sugar*
> *¼ cup dark corn syrup*
> *10 cloves*
> *1½ teaspoons cinnamon*
> *1 teaspoon salt*

Cut the pumpkin into 1-inch cubes.

Combine pumpkin with the vinegar, sugar, corn syrup, cloves, cinnamon and salt in a saucepan. Bring to a boil and cook over low heat until pumpkin looks translucent or until it is easily pricked with a toothpick. Pack into two one-pint sterile jars and seal. Refrigerated, keeps 3 months.

Dill Pickles

> *30 cucumbers*
> *½ cup coarse salt*
> *2 quarts water*
> *2 tablespoons vinegar*
> *4 cloves garlic*
> *4 bay leaves*
> *¼ teaspoon mustard seed*
> *1 teaspoon mixed pickling spice*
> *10 sprigs dill*

Buy even-sized cucumbers and be sure they are very firm. Wash and dry them. Arrange the cucumbers in a crock or jar.

Bring the salt and water to a boil. Cool; then add the vinegar, garlic, bay leaves, mustard seed and pickling spice. Pour over the cucumbers. Arrange the dill over all. The liquid should completely cover the cucumbers; if not, add more salted water. If crock is used, cover with a plate or wooden board to weight it down. Loosely cover with cheesecloth. Keep in a cool place for a week. If you like very green pickles, you might test one at the end of 5 days.

PANCAKES AND BREADS

and Passover Dishes

Cheese Pancakes

    2 egg yolks
    ½ teaspoon salt
    2 tablespoons sugar
    2 cups cottage cheese, drained
    1 cup sifted flour
    2 egg whites, stiffly beaten
    Butter for frying

Beat the egg yolks, salt and sugar together. Stir in the cottage cheese and flour and fold in the egg whites.

    Heat enough butter to cover the bottom of a skillet about ¼ inch. Drop the batter into it by the tablespoon. Fry until browned on both sides. Serve hot, garnished with sour cream and jelly. Serves 4.

Potato Latkes

2 eggs
3 cups grated, drained potatoes
4 tablespoons grated onion
1 teaspoon salt
¼ teaspoon pepper
2 tablespoons cracker or matzo meal
½ cup fat or butter

Beat the eggs and add the potatoes, onion, salt, pepper and meal.

Heat half the fat or butter in a frying pan and drop the potato mixture into it by the tablespoon. Fry until browned on both sides. Keep pancakes hot until all are fried and add more fat or butter as required. Serves 8.

Buckwheat Pancakes

1 cake or package yeast
1 tablespoon sugar
¼ cup lukewarm water
2½ cups lukewarm stock or milk
2 cups buckwheat flour
4 eggs
1 teaspoon salt
3 tablespoons chicken fat or butter, melted
Fat or butter for frying

Combine the yeast, sugar and water. Let stand 5 minutes, then stir in 1 cup stock or milk and 1 cup buckwheat flour. Cover and let rise in a warm place until it bubbles and is double in bulk.

Beat the eggs and salt until thick. Stir in the melted fat or butter, remaining stock or milk, remaining flour and the yeast mixture.

Heat some fat or butter in a skillet and drop the batter into it by the tablespoon. Fry until lightly browned on both sides.

Serve with gravy or with melted butter and sour cream. Serves 6.

Matzo Pancakes

> 3 matzos
> 3 eggs
> 1 teaspoon salt
> 4 tablespoons melted butter or fat
> ⅓ cup matzo meal
> Butter or fat for frying

Soak the matzos in cold water until soft. Drain well and crush into a paste.

Beat the eggs and salt together. Stir in the butter or fat, the soaked matzo and matzo meal. Drop by the tablespoon into hot butter or fat and fry until browned on both sides. Serve with sour cream or stewed fruit, if prepared with butter. Serve with meat dishes if prepared with fat. Serves 2–4.

Matzo-Meal Pancakes (Passover)

> 3 egg yolks
> ½ teaspoon salt
> ½ cup cold water
> ¾ cup matzo meal
> 3 egg whites, stiffly beaten
> Fat for deep frying

Beat together the egg yolks, salt and water. Stir in the matzo meal and fold in the egg whites.

Heat the fat to 375° and drop the batter into it by the tablespoon. Fry until browned on both sides. Drain. Serve with meat dishes or sprinkled with sugar and cinnamon. Serves 2–4.

Potato Muffins

>   2 egg yolks
>   3 cups grated, drained potatoes
>   4 tablespoons grated onion
>   ½ cup sifted flour
>   1 teaspoon salt
>   ½ teaspoon baking powder
>   3 tablespoons melted butter or chicken fat
>   2 egg whites, stiffly beaten

Beat the egg yolks, then stir in the potatoes, onion, flour, salt, baking powder and melted butter or fat. Fold in the egg whites. This quantity makes 12 2-inch muffins.

Grease muffin tin and fill ⅔ full with the mixture. Bake in a 400° oven 25 minutes or until browned. Serve hot.

Challah (Egg Bread)

>   1 cake or package yeast
>   2 teaspoons sugar
>   1¼ cups lukewarm water
>   4½ cups sifted flour
>   2 teaspoons salt
>   2 eggs
>   2 tablespoons salad oil
>   1 egg yolk
>   4 tablespoons poppy seeds

Combine the yeast, sugar and ¼ cup lukewarm water. Let stand 5 minutes.

Sift the flour and salt into a bowl. Make a well in the center and drop the eggs, oil, remaining water and the yeast mixture into it. Work into the flour. Knead on a floured surface until smooth and elastic. Place in a bowl and brush the top with a little oil. Cover with a towel, set in a warm place and let rise 1 hour. Punch down, cover again and let rise until double in bulk. Divide the dough into three equal parts. Between lightly floured hands roll the dough into three strips of even length. Braid them together and place in a baking pan. Cover with a towel and let rise until double in bulk. Brush with the egg yolk and sprinkle with the poppy seeds.

Bake in a 375° oven 50 minutes or until browned.

Makes 1 very large challah. If you wish, divide the dough in six parts and make two large loaves, or make one loaf and many small rolls. You may also bake the bread in a loaf pan.

Note: ⅛ teaspoon saffron can be dissolved in the water if you like additional flavor and color.

Matzo Brie

> 2 *eggs*
> ½ *teaspoon salt*
> 2 *teaspoons grated onion (optional)*
> 2 *matzos*
> *Butter or fat for frying*

Beat the eggs, salt and onion together. Hold the matzos under running water, drain. Then crumble them into the eggs. Mix well.

Heat the butter or fat in a 9-inch skillet and turn the matzo mixture into it. Fry until lightly browned on both sides. Serves 1 or 2.

Matzo Fry

4 whole matzos
3 eggs
½ teaspoon salt
4 tablespoons butter or fat

Hold the matzos under cold running water until softened. Place in a shallow dish. Beat the eggs and salt together and pour over the matzos. Let soak 10 minutes, then fry in the butter or fat until browned on both sides. Serves 2–4.

Matzo Rolls (Passover)

3 eggs
½ teaspoon salt
¾ cup chicken broth or milk
1 cup matzo meal
⅓ cup cake meal
4 tablespoons melted chicken fat or butter

Beat together the eggs and salt. Stir in the broth (or milk) alternately with the matzo meal and cake meal. Add the melted fat or butter. Fill 12 greased muffin cups ⅔ full.

Bake in a 350° oven 30 minutes or until browned.

Potato Chremsel

4 potatoes (1½ pounds), cooked and peeled
2 egg yolks
1 teaspoon salt

⅛ teaspoon pepper
1 tablespoon potato starch
2 egg whites, stiffly beaten
Butter or fat for frying

Mash the potatoes and beat in the egg yolks, salt, pepper and potato starch. Fold in the egg whites.

Melt the butter or fat in a skillet and drop the potato mixture into it by the tablespoon. Fry until browned on both sides. Serve with sour cream or applesauce. Serves 6.

Crescent Rolls

1 cake or package of yeast
¼ cup lukewarm water
¾ cup milk
½ cup butter
4 tablespoons sugar
1 teaspoon salt
3 eggs, beaten
4½ cups sifted flour

Soften the yeast in the water. Scald the milk and stir in the butter, sugar and salt. Cool to lukewarm. Stir in the yeast, eggs and flour. Knead until a smooth dough is formed. Place in a greased bowl and cover. Let rise until doubled in bulk.

Divide dough in thirds and roll each piece into a 10-inch circle. Cut into 12 wedge-shaped pieces. Roll up each piece from the wide end to the point. Place on a greased baking pan and pull ends inward to make a crescent. Let rise until doubled in bulk. Brush with melted butter.

Bake in a 400° oven 15 minutes or until browned. Makes about 36.

French Toast

> 2 eggs
> ½ teaspoon salt
> ¼ cup light cream
> 1 tablespoon sugar
> 4 slices white bread
> 4 tablespoons butter

Beat together the eggs, salt, cream and sugar. Dip the bread in the mixture.

Fry in the butter until browned on both sides. Serve with jelly, syrup or cinnamon-sugar. Serves 2.

DESSERTS

The fragrance of the kitchen before a Jewish holiday is one of the pleasantest memories of childhood. A holiday without cakes and cookies would have been unthinkable. Strudels and honey cake were the overwhelming favorites, but every family had their own special favorites. In point of fact, Jewish cuisine does not place undue emphasis upon sweet desserts, and many meals ended with simple preparations such as fruit compote or plain sponge cake. Rich concoctions involving cream and butter, of course, could not be served after a meat meal and this explains their general absence from Jewish cuisine.

When servings are not specified in this chapter, your family's appetites will determine the number of helpings.

Apple Fritters

> 12 thin slices of peeled apple
> 5 tablespoons sugar
> 2 tablespoons brandy
> 1½ cups sifted flour
> ½ teaspoon salt
> 1 teaspoon baking powder
> 1 egg
> 1 cup milk
> 2 tablespoons melted butter
> Butter for frying

Sprinkle the apple slices with 4 tablespoons of the sugar and the brandy. Set aside.

Sift the flour, salt and baking powder into a bowl. Beat the egg, milk and melted butter together and add to the flour mixture gradually, beating until smooth.

Heat a little butter in a large skillet. The next operation must be done quickly. Pour about 1 tablespoon of the batter into the pan, place an apple slice over it and cover with more batter. Repeat until all the batter and apple slices are used up. Brown on both sides. Makes 12.

Apple Cake

> 1 cup sifted flour
> ½ teaspoon salt
> 1½ teaspoons baking powder
> 6 tablespoons sugar

¼ *pound butter*
1 egg
¼ *cup milk*
4 cups peeled, sliced apples
1 teaspoon cinnamon
½ *cup currant jelly (optional)*

Sift the flour, salt, baking powder and 3 tablespoons sugar into a bowl. Work in half of the butter. Beat the egg and milk together and add, mixing until a dough is formed. Pat onto a buttered 12 x 8-inch baking pan. Arrange the apples in rows. Sprinkle with the cinnamon and remaining sugar. Melt the remaining butter and pour over the top.

Bake in a 400° oven 35 minutes or until apples are tender. Brush with melted jelly if you want to. Cut into squares.

Wonder Cake

½ *cup shortening*
1 cup sugar
5 egg yolks
2¼ *cups sifted cake flour*
¾ *teaspoon salt*
2 teaspoons baking powder
¾ *cup milk*
1 teaspoon vanilla extract
Confectioners' sugar

Cream the shortening and beat in the sugar. Add one yolk at a time, beating after each addition. Sift together the flour, salt and baking powder and add to batter, alternating with the milk. Beat in the vanilla.

Grease a 10-inch loaf pan and dust it lightly with flour. Pour the batter into it.

Bake in a 350° oven 1 hour or until a cake tester comes out clean. Cool on a cake rack. Sprinkle with confectioners' sugar.

Layer Cake

6 egg yolks
⅓ cup sugar
½ cup sifted cake flour
⅛ teaspoon salt
½ teaspoon baking powder
1 teaspoon vanilla
6 egg whites, stiffly beaten

Beat the egg yolks; gradually add the sugar, beating until thick and light in color. Sift flour, salt and baking powder into egg-yolk mixture and fold in. Stir in the vanilla and fold in the egg whites. Turn into 2 8-inch layer-cake pans that have been greased and dusted with flour.

Bake in a 350° oven 20 minutes, or until browned and free from sides of pan. Cool on a cake rack.

Spread whipped cream or cooked fruit or icing or any filling of your choice between the layers of cake.

Sponge Cake

6 egg yolks
1¼ cups sugar
2 teaspoons lemon juice
1 teaspoon grated lemon rind
6 egg whites
¼ teaspoon salt
1½ cups sifted cake flour

Beat the egg yolks; gradually add the sugar, beating until thick and light in color. Stir in the lemon juice and rind.

Beat the egg whites and salt until stiff but not dry. Pile on top of the egg-yolk mixture. Sift the flour over the egg whites and fold in carefully. Turn into a 10-inch tube pan.

Bake in a 325° oven 50 minutes or until browned and free from sides of pan. Invert and let cool.

Butter Cake

> ½ *pound sweet butter*
> *1 cup sugar*
> *6 egg yolks*
> *1 cup sifted flour*
> *1 cup sifted cornstarch*
> *1 teaspoon baking powder*
> ⅛ *teaspoon salt*
> *2 tablespoons brandy*
> *6 egg whites, stiffly beaten*
> *Confectioners' sugar*

Cream the butter; gradually add the sugar, beating until light and fluffy. Add 1 egg yolk at a time, beating after each addition. Sift the flour, cornstarch, baking powder and salt into butter mixture; mix lightly and stir in the brandy. Fold in the egg whites. Turn into a buttered 9-inch spring-form pan.

Bake in a 325° oven 1 hour or until a cake tester comes out clean. Cool on a cake rack. Dust with confectioners' sugar.

Mandelbrot

1¼ cups sifted flour
¼ teaspoon salt
1 teaspoon baking powder
4 eggs
1 cup sugar
3 tablespoons salad oil
1 teaspoon vanilla extract
1 tablespoon cinnamon
1 cup almonds

The finished cake should be only 2½ inches high so use 2 9-inch loaf pans for baking.

Sift together the flour, salt and baking powder.

Beat the eggs until thick; gradually add the sugar, beating until lemon-colored. Stir in the oil and vanilla. Gradually add the flour mixture, then the almonds.

Cover the bottoms of the 2 greased loaf pans with a little of the batter. Sprinkle with a little cinnamon and repeat the steps until all the batter and cinnamon is used up.

Bake in a 350° oven 35 minutes or until a cake tester comes out clean. Cool, then cut in ½-inch slices. Arrange slices on a baking sheet and lightly brown in a 400° oven.

Honey Cake

3½ cups sifted flour
¼ teaspoon salt
1½ teaspoons baking powder
1 teaspoon baking soda

½ teaspoon cinnamon ⎫
¼ teaspoon nutmeg ⎪
⅛ teaspoon powdered cloves ⎬ *(optional)*
½ teaspoon ginger ⎭
4 eggs
¾ cup sugar
4 tablespoons salad oil
2 cups dark honey
½ cup brewed coffee
1½ cups nuts, walnuts or almonds

Sift the flour, salt, baking powder, baking soda, cinnamon, nutmeg, cloves and ginger together. (The spices depend on your personal taste, so add or not, as you prefer.)

Beat the eggs, gradually adding the sugar. Beat until thick and light in color. Beat in the oil, honey and coffee; stir in the flour mixture and the nuts.

Oil an 11 x 16 x 4-inch baking pan and line it with aluminum foil. Or, if you want 2 smaller cakes, use 2 9-inch loaf pans. Turn the batter into the pan or pans.

Bake in a 325° oven, 1¼ hours for the large cake, 50 minutes for the 2 smaller ones, or until browned and a cake tester comes out clean. Cool on a cake rack before removing from pan.

Coffee Cake

1¼ cups sifted flour
¼ teaspoon salt
½ teaspoon baking soda
1 teaspoon cream of tartar
4 tablespoons butter
1 cup sugar
1 egg

½ cup milk
⅓ cup brown sugar
2 tablespoons cinnamon

Sift together the flour, salt, baking soda and cream of tartar.

Cream the butter, gradually adding the sugar. Beat until fluffy, then add the egg. Add the flour mixture, alternating with the milk. Turn into an 8-inch-square buttered pan. Sprinkle with the brown sugar and cinnamon.

Bake in a 350° oven 30 minutes. Serve warm.

Yeast Crumb Cake

1 cake or package yeast
⅓ cup sugar
2 tablespoons lukewarm water
⅓ cup lukewarm milk
2 cups sifted flour
4 tablespoons butter
2 eggs
¼ teaspoon salt
3 tablespoons melted butter

Combine the yeast, 1 tablespoon sugar and the water in a small bowl. Set aside for 5 minutes. Stir in the milk and ½ cup flour. Cover and set aside to rise in a warm place until double in bulk.

Cream the butter, gradually adding the remaining sugar, the eggs and salt. Add the yeast mixture and remaining flour, then knead until smooth. Turn out into an 8-inch-square pan and brush with the butter. Cover and let rise until double in bulk. Combine the following ingredients and sprinkle over the top:

4 tablespoons melted butter
4 tablespoons sugar
2 teaspoons cinnamon
1 cup sifted flour

Bake in a 350° oven 30 minutes or until brown.

Carrot Cake

9 egg yolks
1¾ cups sugar
1½ cups finely mashed cooked carrots
1 tablespoon grated orange rind
1 tablespoon brandy
2½ cups ground almonds
9 egg whites, stiffly beaten

Beat the egg yolks; gradually add the sugar, beating until thick. Stir in the carrots, orange rind, brandy and almonds. Fold in the egg whites thoroughly. Turn into a greased 9-inch spring form.

Bake in a 325° oven 50 minutes. Cool and remove sides of pan.

Date-Nut Torte

2 tablespoons bread crumbs
8 egg yolks
1¼ cups sugar
2 cups chopped walnuts or pecans
2 cups chopped dates
4 tablespoons flour
8 egg whites, stiffly beaten

Grease a 9-inch spring-form pan and dust it with the bread crumbs.

Beat the egg yolks; gradually add the sugar. Beat until thick. Mix together the nuts, dates and flour. Stir into the egg-yolk mixture. Fold in the egg whites thoroughly. Turn into the spring form.

Bake in a 325° oven 40 minutes. Cool and remove the sides of the pan.

Mocha Torte

> 1 cup sifted cake flour
> ⅛ teaspoon salt
> 1 teaspoon baking powder
> 4 egg yolks
> 1 cup powdered sugar
> 4 tablespoons ice water
> 1 tablespoon coffee essence
> 1 teaspoon vanilla
> 4 egg whites, stiffly beaten

Sift the flour, salt, and baking powder 3 times. Beat the egg yolks; gradually add the sugar, beating until thick and light in color. Combine water, coffee essence and vanilla. Add to yolk mixture alternately with the flour mixture. Fold in the egg whites. Turn into 2 8-inch layer-cake pans that have been greased and dusted with flour.

Bake in a 350° oven 20 minutes or until a cake tester comes out clean. Cool. Spread one layer with following mixture:

> 3 cups whipped cream
> 2 tablespoons confectioners' sugar
> 1 tablespoon coffee essence

Top with remaining layer.

Fruit Pies

CRUST:

> 2 cups sifted flour
> ¾ teaspoon salt
> ¾ cup shortening
> 6 tablespoons ice water

Sift the flour and salt into a bowl. Add the shortening with a pastry blender or 2 knives until the consistency of coarse sand. Sprinkle the ice water over the mixture and toss until balls of dough are formed. Form into a ball and chill ½ hour. Divide dough in half and roll out as thin as possible. Fit into a 9-inch pie plate and fill. Cover with other piece of dough and seal the edges. Make a few slashes in the top and brush top with beaten egg yolks. Serves 6–8.

FILLINGS FOR FRUIT PIES

RHUBARB:

> 3 cups rhubarb sliced in 1-inch lengths
> 1 cup sugar
> 3 tablespoons flour
> ⅛ teaspoon salt
> 2 tablespoons butter (optional)

Mix the rhubarb, sugar, flour and salt together and fill the pie plate. Dot with the butter if you want to use it. Cover with top crust.
 Bake in a 400° oven 45 minutes or until browned.

APPLE:

> 6 cups peeled, sliced apples
> ¾ cup sugar

⅛ teaspoon salt
2 teaspoons lemon juice
½ teaspoon cinnamon

Mix together all the ingredients. Fill and bake as described above.

BERRIES:

4 cups blueberries, raspberries or blackberries
¾ cup sugar
2 tablespoons flour
2 teaspoons lemon juice
⅛ teaspoon nutmeg

Mix all the ingredients together. Fill and bake as described above.

CHERRY:

2 cans (#2) sour red cherries, drained
1 cup sugar
⅛ teaspoon salt
2 tablespoons flour
¼ teaspoon almond extract

Mix all the ingredients together. Fill and bake as described above.

Pineapple Chiffon Pie

3 tablespoons cornstarch
2 cups pineapple juice
1 cup sugar
⅛ teaspoon salt
2 cups crushed pineapple
4 egg whites
Baked 9-inch pie shell (page 167)

Mix the cornstarch with a little pineapple juice until smooth. Combine in a saucepan with the rest of the juice, ½ cup of the sugar and the salt. Cook over low heat, stirring constantly, until thickened. Stir in the pineapple and cool.

Beat the egg whites until stiff but not dry; then beat in the remaining sugar. Fold into the pineapple mixture and fill the pie shell. Chill. Serves 6–8.

Kichlach (Puffy Egg Cookies)

> *3 eggs*
> *½ cup salad oil*
> *2 tablespoons sugar*
> *1 cup sifted flour*
> *¼ teaspoon salt*
> *4 tablespoons poppy seeds (optional)*

Beat the eggs until light, then beat in the oil, sugar, flour and salt. Beat until very smooth. Stir in the poppy seeds, if you like.

Drop by the teaspoon onto a greased baking sheet, leaving about 3 inches between each. (They spread and puff in baking.)

Bake in a 325° oven 15 minutes or until browned on the edges. Makes approximately 36.

Almond Squares

> *½ pound butter*
> *1 cup sugar*
> *¾ cup sifted flour*
> *1 cup ground almonds*

¼ teaspoon salt
6 egg yolks
16 blanched almonds

Cream the butter; gradually add the sugar, beating until light and fluffy. Combine the flour, almonds and salt. Add a little at a time to the butter mixture, alternating with 1 egg yolk at a time. Turn into a buttered 8-inch-square pan. Mark off 16 squares and place an almond in the center of each square.

Bake in a 350° oven 35 minutes or until browned. Cut into squares.

Honey Cookies

2 eggs
¾ cup sugar
⅔ cup honey
¾ cup sliced almonds
2½ cups sifted flour
½ teaspoon cinnamon
½ teaspoon ginger
½ teaspoon nutmeg
¼ teaspoon baking soda
½ cup chopped candied fruit peel

Beat the eggs and sugar together until fluffy. Beat in the honey and then the nuts. Sift together the flour, cinnamon, ginger, nutmeg and baking soda. Work into the honey mixture and add the candied fruit. Form into a ball and chill.

Roll out as thin as possible and cut into shapes. Place on a greased cooky sheet.

Bake in a 350° oven 15 minutes or until browned. Makes about 36 3-inch cookies.

Macaroons

> 5 egg whites
> 1 cup sugar
> ½ pound almond paste
> ¼ cup white corn meal

Beat the egg whites until stiff but not dry; beat in the sugar. Mix together the almond paste and corn meal and fold in the egg whites. Drop by the teaspoon onto a lightly greased cooky sheet; flatten slightly with a wet knife.

Bake in a 350° oven 20 minutes or until delicately browned. Makes about 6 dozen.

Chocolate Chip Cookies

> ¼ pound butter
> ½ cup granulated sugar
> ¼ cup brown sugar
> 1 egg, beaten
> 1⅛ cups sifted flour
> ½ teaspoon salt
> ½ teaspoon baking soda, in 1 tablespoon water
> 1 teaspoon hot water
> ½ teaspoon vanilla extract
> 1 package (6-ounce) chocolate bits
> ½ cup chopped walnuts

Cream the butter and beat in the granulated and brown sugars until light and fluffy. Beat in the egg. Sift together the flour, salt and baking soda. Work into the butter mixture. Stir in the hot water, vanilla, chocolate bits and nuts. Drop by the teaspoon onto a greased cooky sheet.

Bake in a 375° oven 10 minutes or until browned. Makes about 40.

Paul's Brownies

>   1 cup sifted cake flour
>   ½ cup sifted unsweetened cocoa
>   ¼ pound butter
>   1¼ cups brown sugar, packed
>   2 eggs
>   ⅛ teaspoon salt
>   ½ cup milk
>   1 teaspoon vanilla extract
>   1 cup pecans

Sift the flour and cocoa together. Cream the butter and gradually add the brown sugar. Beat until light and fluffy. Add the eggs and salt, mixing well. Add the cocoa mixture alternately with the milk. Stir in the vanilla and pecans. Pour into a greased 8-inch-square pan.

Bake in a 350° oven 30 minutes or until a cake tester comes out clean. Cut into 1½-inch squares before cooling. Makes about 25.

Hamentaschen

YEAST DOUGH:
>   2 cakes or packages yeast
>   ¼ cup lukewarm water
>   ¾ cup scalded milk, cooled
>   5 cups sifted flour
>   ¾ cup sugar
>   1½ teaspoons salt
>   3 eggs
>   1 cup melted butter

Soften the yeast in the water for 5 mi... the milk. Stir in 2 cups of the flour, the Add one egg at a time, beating after each in the butter and remaining flour. Knead fo... utes, then place in a bowl; cover with a tow... et rise in a warm place until doubled in bulk. Pu... down and knead on a lightly floured board for 5 minutes. Divide the dough in half and roll out to ¼-inch thickness. Cut into 4-inch squares. Place a heaping tablespoon of filling on each and fold the dough over into a triangle, sealing the edges. Let rise until doubled in bulk and brush with beaten egg yolk.

Bake in a 375° oven 25 minutes or until browned. Makes about 3½ dozen.

HONEY DOUGH FOR HAMENTASCHEN:
4 cups sifted flour
½ teaspoon salt
1 teaspoon baking powder
½ cup shortening, softened
4 eggs
1 cup honey

Sift the flour, salt and baking powder into a bowl. Make a well in the center and place the shortening, eggs and honey in it. Work together with the hand until a dough is formed. Roll out and cut into 4-inch squares. Place a heaping tablespoon of filling on each and fold over into a triangle, sealing the edges.

Bake in a 350° oven 20 minutes or until browned. Makes about 3 dozen.

FILLINGS FOR HAMENTASCHEN

POPPY SEED:
2 cups poppy seeds
1 cup milk

¾ *cup honey*
1 *teaspoon grated lemon rind*
½ *cup seedless raisins*

Have the poppy seeds ground or put through your food chopper. Combine with the milk and honey. Cook over low heat, stirring frequently, until thick. Stir in the lemon rind and raisins. Cool and fill the dough.

PRUNE:

2 *cups lackwa (prune butter)*
½ *cup ground almonds*
1 *tablespoon grated orange rind*

Mix ingredients together and fill dough.

Poppy-Seed Candy

1 *pound poppy seeds*
2 *cups honey*
½ *cup sugar*
2 *cups chopped nuts*
½ *teaspoon powdered ginger*

Have the poppy seeds ground for you when you buy them. If this is not possible, grind them in a food chopper or pound with a mortar and pestle.

Cook together the honey and sugar until syrupy. Stir in the poppy seeds and cook until mixture is thick, about 20 minutes. Stir frequently. (Drop a little on a wet surface; if it doesn't run, it is thick enough.) Stir in the nuts and ginger.

Moisten hands; pat out mixture onto wet board to thickness of about ½ inch. Let cool 5 minutes, then cut into diamonds or squares with a sharp knife. When knife sticks, dip into hot water. Cool completely and lift from board with a spatula.

Poppy-Seed Crunch

>   3 tablespoons butter
>   ½ cup sugar
>   2 egg yolks
>   1 teaspoon vanilla extract
>   1½ cups sifted flour
>   ⅛ teaspoon salt
>   1½ teaspoons baking powder
>   2 egg whites
>   1 cup brown sugar
>   1 cup ground poppy seeds

Cream the butter, gradually adding the sugar. Mix in the egg yolks and vanilla. Sift together the flour, salt and baking powder and add to butter mixture. Pat into an 8-inch-square buttered pan. (The mixture won't hold together when you're handling it, but it will when baked.)

Beat the egg whites until they begin to stiffen, then beat in the brown sugar. Fold in the poppy seeds. Spread over the dough.

Bake in a 350° oven 25 minutes or until delicately browned and set. Cool and cut into squares.

Taiglach

>   2½ cups sifted flour
>   ⅛ teaspoon salt
>   1 teaspoon baking powder
>   4 eggs
>   4 tablespoons salad oil
>   1 pound dark honey
>   ¾ cup brown sugar
>   1 teaspoon powdered ginger

½ teaspoon nutmeg
2 cups filberts (or other nuts except peanuts)
½ cup candied cherries (optional)

Sift the flour, salt and baking powder into a bowl. Make a well in the center and drop the eggs and oil into it. Work into the flour and mix until a dough is formed. Break off pieces of dough and roll into pencil-thick strips. Cut into ½-inch pieces and place on a lightly greased cooky sheet.

Bake in a 350° oven 20 minutes or until browned. Shake the pan once or twice. Cool.

Cook the honey, brown sugar, ginger and nutmeg for 15 minutes. Drop the baked dough into it and cook for 5 minutes. Add the nuts and cook 10 minutes additional, stirring frequently. Test the mixture by dropping a little on a wet surface; if it holds together, it's done; if not, cook until it does. Turn out onto a wet board and let cool until easy enough to handle. Then shape into 3-inch balls between moistened hands. Decorate with the candied cherries, if you wish. Makes approximately 36.

Bread Pudding

8 slices stale white bread
3½ cups milk
4 tablespoons butter
½ cup sugar
2 eggs, beaten
½ cup sherry
¼ teaspoon nutmeg
¼ teaspoon cinnamon
½ teaspoon salt
1 cup seedless raisins

Cut the bread into small cubes (there should be about 2 cups). Scald the milk and add the butter; pour over the bread cubes. Let soak for 10 minutes, then blend in the sugar, eggs, sherry, nutmeg, cinnamon, salt and raisins. Pour into a 1½-quart buttered baking dish and place in a pan of hot water.

Bake in a 375° oven 1 hour or until a knife inserted in the center comes out clean. Serves 6.

Rice Pudding

> ½ cup raw rice
> ¾ teaspoon salt
> 2 cups boiling water
> 3½ cups milk
> 3 eggs
> ⅓ cup granulated sugar
> 2 teaspoons vanilla extract
> ½ cup seedless raisins
> 2 tablespoons melted butter
> Nutmeg

Cook the rice in the salted water for 15 minutes. Drain, then add the milk.

Beat the eggs, then add the sugar, vanilla, raisins, butter and rice mixture. Pour into a 2-quart casserole; place in a pan of water.

Bake in a 325° oven 25 minutes. Stir and sprinkle with nutmeg. Reduce the heat to 300° and bake 1 hour longer or until a knife inserted in the center comes out clean. Serve warm or cold. Serves 6–8.

Prune Whip

> 2 jars strained prunes (baby food)
> ½ cup prune juice
> 1 cup boiling water

½ cup brown sugar
¼ teaspoon cinnamon
⅛ teaspoon salt
2 tablespoons cornstarch
¼ cup cold water
1 tablespoon lemon juice
½ cup chopped walnuts (optional)
2 egg whites, stiffly beaten
½ cup heavy cream, whipped

Combine the prunes, prune juice, boiling water, sugar, cinnamon and salt. Mix the cornstarch and cold water until smooth and stir into the prune mixture. Cook over low heat, stirring constantly until mixture reaches the boiling point. Cook 5 minutes longer, stirring occasionally. Cool and stir in the lemon juice and walnuts, if used. Fold in the egg whites and whipped cream. Pour into a mold or 8 individual serving dishes. Chill and garnish with a dab of whipped or sour cream. Serves 8.

Potato-Flour Sponge Cake (Passover)

7 egg yolks
2 eggs
1¾ cups powdered sugar
2 teaspoons lemon juice
1 teaspoon lemon rind
7 egg whites
⅞ cup sifted potato flour
⅛ teaspoon salt

Beat the egg yolks and eggs; gradually add the sugar, beating until thick and light in color. Stir in the lemon juice and rind.

Beat the egg whites until stiff but not dry. Pile onto the sugar mixture. Sift the potato flour and salt over it and fold in gently. Turn into a 10-inch tube pan.

Bake in a 350° oven 45 minutes, or until lightly browned and the sides shrink from the pan. Invert and cool.

Macaroons (Passover)

> 6 egg whites
> 1 cup sugar
> ½ pound Passover almond paste
> ¼ cup matzo meal

Beat the egg whites until peaks begin to form; beat in the sugar and fold in the almond paste and matzo meal. Drop by the teaspoon onto a greased baking pan and flatten slightly with a wet knife.

Bake in a 350° oven 20 minutes or until delicately browned. Makes about 6 dozen.

Banana Cake (Passover)

> 7 egg yolks
> 1 cup sugar
> ¼ teaspoon salt
> 1 cup mashed bananas
> ¾ cup sifted potato starch
> 1 cup coarsely chopped walnuts
> 7 egg whites, stiffly beaten

Beat the egg yolks until thick. Add the sugar and salt and beat until fluffy and lemon colored. Stir in the bananas and potato starch, then the walnuts. Fold in the egg whites. Pour into 2 greased 9-inch layer-cake pans.

Bake in a 350° oven 30 minutes or until a toothpick comes out clean. Cool on a cake rack. Spread whipped cream and sliced bananas on one layer and cover with the other. Serves 6–8.

Fluffy Walnut Cake (Passover)

>   ¾ *cup matzo meal*
>   ¾ *cup potato starch*
>   ½ *teaspoon salt*
>   6 *egg yolks*
>   1¾ *cups sugar*
>   1 *cup orange juice*
>   1½ *cups ground walnuts*
>   1 *tablespoon grated lemon rind*
>   6 *egg whites, stiffly beaten*

Mix the matzo meal, potato starch and salt together.

Beat the egg yolks until thick; gradually add the sugar, beating until lemon colored. Add the matzo-meal mixture alternately with the orange juice. Fold in the walnuts and lemon rind, and then the egg whites. Turn into a 9-inch tube pan.

Bake in a 325° oven 1 hour or until browned and the cake shrinks away from the sides of the pan. Cool on a cake rack. Serves 6–8.

STRUDEL

Strudel dough takes practice and patience. Your first few tries may not be successful but try again—it's worth it.

Stretched Strudel Dough

> 3 cups sifted flour
> ¼ teaspoon salt
> 2 eggs
> 3 tablespoons salad oil
> ¼ cup lukewarm water

Sift the flour and salt into a bowl. Make a well in the center and drop the eggs, oil and water into it. Work into the flour, mixing until the dough leaves the sides of the bowl. Knead the dough for 10 minutes or until very smooth and elastic. Place a warm bowl over it and let it rest for 20 minutes.

For the next step you will need a large working surface—one you can walk around. A kitchen table is best. Cover it with a cloth and sprinkle with flour. Roll out the dough as thin as you can. Now you must begin stretching it. Flour the knuckles of your hands and gently pull the dough toward you from underneath, using the back of your hands. Change your position as the dough stretches so as not to put too much strain on any one part. Stretch until the dough is transparent, then brush with oil or melted butter. Cut away any thick edges. Spread one of the following fillings over half the surface; then raise the cloth and roll up the strudel from the filled side, guiding it with the other hand. Place on a heavily greased baking pan. Brush with oil or melted butter.

Bake in a 400° oven 35 minutes or until browned and crisp. Cut into 1½-inch slices immediately. Makes about 40 slices.

If you like a smaller strudel, divide dough in two before rolling, and make 2 strudels.

Unstretched Strudel Dough

> 2½ cups sifted flour
> ½ teaspoon salt
> 1 teaspoon baking powder
> 1 egg
> ⅔ cup ice water
> 4 tablespoons salad oil

Sift the flour, salt and baking powder into a bowl. Make a well in the center and drop the egg, water and oil into it. Work into the flour and knead until smooth and elastic. Place a warm bowl over it for 30 minutes.

Roll out as thin as possible and spread with one of the fillings. Roll up like a jelly roll, brush with oil or melted butter. Bake in a 350° oven 45 minutes. Cut into 1-inch slices while hot. Makes about 36.

FILLINGS FOR STRUDEL

CABBAGE:

6 cups finely shredded cabbage
½ cup minced onions
⅓ cup butter or chicken fat
1½ teaspoons salt
¼ teaspoon pepper
2 teaspoons sugar

Cook the cabbage and onions in the butter or fat over low heat 25 minutes, mixing occasionally. Stir in the salt, pepper and sugar. Cool. Roll up in strudel dough. Serve with main courses.

LIVER:

1½ pounds calf's liver
1 cup diced onions
½ cup chicken fat
1½ teaspoons salt
¼ teaspoon pepper

Cube the liver and cook it with the onions in the chicken fat until the liver loses its redness. Chop the liver (use the fat remaining in the pan, too). Add the salt and pepper. Spread on the oiled strudel dough and roll up. Make 4 small strudels with this filling.

SAUERKRAUT:

> 2 pounds sauerkraut
> ¼ cup minced onion
> 4 tablespoons butter or chicken fat
> 2 teaspoons sugar
> ¼ teaspoon pepper

Rinse the sauerkraut in cold water and drain. Cook the sauerkraut and onions in the butter or fat for 15 minutes. Stir in the sugar and pepper. Cool. Spread on strudel dough and roll up. Serve with main courses.

APPLE:

> 1 cup fine bread crumbs
> 1½ cups ground nuts
> 4 cups chopped apples
> 2 tablespoons grated lemon rind
> 2 tablespoons lemon juice
> 1 cup seedless raisins
> ½ cup sugar
> 2 teaspoons cinnamon

Sprinkle bread crumbs over half the oiled, stretched dough. Sprinkle the nuts over it and spread evenly with the hand. Mix the apples, lemon rind, lemon juice and raisins together. Spread over the nuts. Sprinkle with the sugar mixed with the cinnamon. Roll up.

MIXED DRIED FRUIT:

> 1 pound prunes, pitted
> 1 pound dates, pitted
> 1 cup candied mixed fruit
> 1 whole lemon
> 2 cups walnuts
> 4 tablespoons sugar
> 2 teaspoons cinnamon

Grind the prunes, dates, candied fruit, lemon and walnuts in a food chopper. Divide the strudel dough in two. Spread the rolled dough with oil and use half the fruit mixture for each. Roll up, brush with oil and sprinkle with a mixture of sugar and cinnamon.

CHERRY:

> *1½ cups finely ground nuts (except peanuts)*
> *4 cups canned sour red cherries, pitted and drained*
> *1 cup sugar*

Spread the nuts over half the oiled strudel dough and cover with the cherries. Sprinkle with the sugar. Roll up.

POPPY SEED:

> *1 pound ground poppy seeds*
> *¾ cup honey*
> *½ cup light cream*
> *½ cup currants*
> *1 tablespoon grated lemon rind*
> *3 tablespoons melted butter*

Cook the poppy seeds, honey, cream and currants until thick. Stir in the lemon rind and cool. Spread on the oiled strudel dough and roll up. Brush with the butter.

WALNUT:

> *4 cups chopped walnuts*
> *2 cups chopped apples*
> *1 cup sugar*
> *2 teaspoons cinnamon*
> *3 tablespoons grated lemon rind*

Spread the nuts on half the oiled dough and cover with the apples. Sprinkle with the sugar mixed with the cinnamon and lemon rind. Roll up.

This strudel is also good made in thin rolls, so divide the dough into four, if you like, and make 4 small strudels.

COCONUT:

>   3 *cups shredded coconut*
>   2 *cups ground nuts*
>   2 *cups fruit preserves*

For this recipe (because coconut is very rich), it is advisable to make 4 small strudels. Mix the coconut, nuts and preserves together and spread ¼ amount on each piece of oiled dough. Roll up.

CHEESE:

>   2 *eggs*
>   ½ *cup sugar*
>   ½ *pound pot cheese*
>   ½ *pound cream cheese*
>   1 *teaspoon vanilla extract*
>   3 *tablespoons dry bread crumbs*

Beat the eggs and sugar until thick. Add the pot cheese, cream cheese and vanilla. Continue to beat until smooth.

Sprinkle the bread crumbs over half the oiled strudel dough and spread the cheese mixture over it. Roll up.

Some Suggested Menus

Dairy Meals

Chopped eggs and onions
Dairy potato soup
Baked salmon
Baked potato
Mocha torte
Coffee or tea

Jennie's herring salad
Cheese blintzes with sour cream
Yeast crumb cake
Coffee or tea

Dairy split pea soup
Fried fish
Mushrooms in sour cream
Apple fritters
Coffee or tea

Cold borsch
Lemon fish
Sweet-and-sour red cabbage
Noodle kugel
Coffee or tea

Baked herring and potatoes
Rice-stuffed cabbage
Butter cake
Coffee or tea

Meat Meals

Lentil soup
Roast tongue
Roast potato
Prunes and beets
Cookies
Tea or black coffee

Barley-bean soup
Baked veal chops
Sweet-potato tzimmes
Fried eggplant
Apple strudel
Tea or black coffee

Flanken soup
Boiled fllanken
Boiled potato
Horseradish
Grapefruit or melon
Tea or black coffee

Herring forshmak
Noodle-stuffed duck
Tzibbele kugel
Applesauce
Sponge cake
Tea or black coffee

Tomato soup
Stuffed breast of veal
Potato stuffing
Honeyed carrots
Mandelbrot
Tea or black coffee

Friday Night Dinner

Gefilte fish
Chicken soup with mandlen
Baked chicken
Potato kugel
Carrot-and-apple tzimmes
Honey cake
Tea or black coffee

Saturday Lunch

Chopped liver
Meat cholent
Cucumber salad
Cookies
Tea or black coffee

Passover Meals

CEREMONIAL PASSOVER FOODS

Boiled carp with horseradish
Roast turkey
Mixed-fruit tzimmes
Beet salad
Potato-flour sponge cake
Tea or black coffee

Passover borsch
Matzo brie
Applesauce
Tea or coffee

Chopped eggs and onions
Chremsel and sour cream
Fluffy walnut cake
Tea or coffee

Suggested Menus for the Principal Holidays

Rosh Hashonah (Jewish New Year)

Fresh sliced pineapple
Chicken soup with kreplach
Roast turkey
Sweet potato royal
Bean panache
Cucumber salad
Cranberry sauce
Honey cake
Taiglach
Challah
Assorted nuts
Tea or black coffee

Gefilte fish with beet horseradish
Chicken soup with liver balls
Roast beef
Kishke
Cauliflower
Wilted red-cabbage slaw
Apple cake
Taiglach
Dates and figs
Challah
Tea or black coffee

Succoth (Feast of the Tabernacles)

> *Chilled tomato juice*
> *Gefilte fish with beet horseradish*
> *Barley-bean soup*
> *Roast stuffed goose*
> *Noodle-apple pudding*
> *Braised kale*
> *Pickled vegetable salad*
> *Pineapple chiffon pie*
> *Challah*
> *Assorted nuts, dates and figs*
> *Tea or black coffee*

> *Cranshaw melon*
> *Celery and olives*
> *Gefilte fish with horseradish*
> *Potage à la Reine*
> *Cornish hen with wild rice*
> *Cauliflower*
> *Brussels sprouts*
> *Beet salad*
> *Mandelbrot*
> *Challah*
> *Assorted fresh fruits*
> *Tea or black coffee*

Chanukah (Festival of Lights)

Chanukah is one of the most joyous Jewish holidays. It is a holiday which is celebrated by lighting of the Chanukah candles, giving of gifts to children, spinning the *dreidel*, and it is one that brings out the joy of family living.

In the way of food, potato *latkes* (pancakes) and *kreplach* are the order of the day. Because this is an eight-day holiday, we are not giving you any single menu but suggest experimenting with a variety of dishes suitable for festive occasions in the winter. Traditionally, the custom was to consume vast quantities of potato pancakes while the elders and children had fun spinning the *dreidel*.

Purim (Festival of Lots)

Chopped eggs and onions
Cabbage borsch
Stuffed breast of veal
Leaf spinach
Mixed-fruit tzimmes
Dill pickles
Hamentaschen
Assorted nuts
Tea or black coffee

Petcha
Chicken soup with mandlen
Baked chicken
Noodle kugel
Green peas
Cole slaw
Stewed prune compote
Hamentaschen
Dates and figs
Tea or black coffee

Pesach (Festival of Passover)

See also SOME SUGGESTED MENUS, PAGE 187

> *Half grapefruit*
> *Celery and olives*
> *Gefilte fish with beet horseradish*
> *Meat borsch*
> *Roast pullet*
> *Fresh asparagus*
> *Potato kugel*
> *Hearts of lettuce with Russian dressing*
> *Passover macaroons*
> *Assorted nuts*
> *Matzos*
> *Tea or black coffee*

> *Spanish melon*
> *Chicken soup with knaidlach*
> *Pot roast with vegetables*
> *Matzo farfel*
> *Honeyed carrots*
> *Cole slaw*
> *Sponge cake*
> *Matzos*
> *Fresh fruit bowl*
> *Tea or black coffee*

Shavuoth (Festival of the Torah)

(DAIRY DISHES ARE FAVORED DURING THIS TWO-DAY HOLIDAY)

> *Cantaloupe*
> *Vegetarian chopped liver*

SHAVUOTH (*continued*)

> Scotch barley soup
> Baked stuffed brook trout Meunière
> Asparagus
> Fresh baby lima beans
> Tossed salad with French dressing
> Blueberry pie
> Assorted nuts
> Tea or coffee

> Fresh grapefruit cup
> Split pea soup
> Sweet-and-sour fish
> Creamed beets
> Mushroom piroshki
> Cucumber salad
> Date-nut torte
> Fresh fruit bowl
> Tea or coffee

Traditional Friday Night

> Fresh fruit cocktail
> Gefilte fish with beet horseradish
> Chicken soup with noodles
> Pot-roasted chicken
> Sweet-potato-and-prune tzimmes
> String beans
> Hearts of lettuce and tomato salad
> Applesauce
> Almond squares
> Challah
> Assorted nuts
> Tea or black coffee

TRADITIONAL FRIDAY NIGHT (*continued*)

Chopped liver
Boiled carp
Flanken soup
Boiled flanken
Carrots and peas
Potato cholent
Mixed green salad with French dressing
Fruit compote
Apple strudel
Dates and figs
Challah
Tea or black coffee

Index

Index